Praise for *Breakthrough*

'William Pao's book vividly brings to life the people involved in medical research and development to find new ways to advance clinical outcomes for patients. The perseverance; intuition; iterations required to engineer that "sweet spot" are narrated in a very readable way.'

William 'Bill' Burns, Former CEO of Roche Pharma

'*Breakthrough* reads like a mystery novel. It draws the reader in, first by describing the scientific discovery that provides key insight into a disease and then by weaving a tapestry of those who overcame almost unimaginable challenges to design and test the medicines we have today that cure, prevent, or vastly improve outcomes for those stricken with grievous illnesses... inspirational; a must read for anyone interested in or associated with drug discovery and development.'

Mace Rothenberg MD, President and Executive Director of Museum of Medicine and Biomedical Discovery

'An essential read for anyone interested in how basic science is translated for human benefit. *Breakthrough: The Quest for Life-Changing Medicines* delves into the fascinating and often surprising journey of drug discovery, unravelling the fascinating biology and convoluted research paths that lead to the development of new game-changing therapies. Authored by an acclaimed doctor, cancer scientist and drug developer; this book tells the captivating stories that bring the world of biomedical research and drug discovery to life.'

Dr Norman E. Sharpless, Former Director of the National Cancer Institute and Former Acting Commissioner of the FDA

'Dr William Pao has been a pivotal force in the field of cancer research for decades, channelling the tragedy of his father's untimely death from cancer into a passion for scientific discovery. In *Breakthrough*, Dr Pao uses eight interesting vignettes to demonstrate that the process of discovery is complex, but that it exemplifies human ingenuity and the extraordinary power of team science.'

Margaret Foti, PhD, MD, CEO of American Association for Cancer Research

'A truly enlightening journey through many of the most important medical advances of the last few decades. The science underlying the discoveries blends seamlessly with personal insights into the researchers, physicians and patients who brought the work to life. All written in a style that will engage the curious lay person to those of us who were fortunate to witness some of these miracles firsthand.'

Howard A. 'Skip' Burris III, MD,
President, Sarah Cannon Research Institute (SCRI)
and Past President, American Society of Clinical Oncology (ASCO)

'William Pao presents eight stories of biomedical achievement in a compelling, captivating, and remarkably complete fashion. These stories – and Pao's own insights as both an accomplished physician-scientist and successful drug developer – capture the exhilarating process of discovery and the many individuals and teams that collectively contribute to the launching of successful medicines. *Breakthrough: The Quest for Life-Changing Medicines* is not just about chronicling a series of success stories; it is also about providing hope for much more of the same in the decades ahead.'

Tyler Jacks, Founding Director,
Koch Institute for Integrative Cancer Research;
President, Break *Through* Cancer

'Pao traces seminal innovations from idea through challenge, setback, and risk, harnessing determination and luck to bring them to fruition. *Breakthrough* shares the excitement of bold discovery that captivates scientific researchers – motivated by the unyielding quest to save lives.'

Martha Liggett, Executive Director,
American Society of Haematology

BREAKTHROUGH

The Quest for Life-Changing Medicines

WILLIAM PAO, MD PhD

A Oneworld Book

First published in the United Kingdom, United States of America,
Republic of Ireland and Australia by Oneworld Publications Ltd, 2024

A CIP record for this title is available from the British Library

ISBN 978-0-86154-734-0
eISBN 978-0-86154-735-7

Illustrations by Merlyn Harvey
Typeset by Hewer Text UK Ltd, Edinburgh
Printed and bound in Great Britain by Clays Ltd, Elcograf S.p.A.

Oneworld Publications Ltd
10 Bloomsbury Street
London WC1B 3SR
England

To my dad, who died too young

CONTENTS

FOREWORD
BY HAROLD VARMUS

THE DISTRESSING SYMPTOMS AND LETHAL CONSEQUENCES of disease are unwelcome commonalities of the human condition. Throughout history, various approaches – influenced by culture, religion, geography and social structures – have been pursued in efforts to ward off or reverse disease. Largely due to recent advances in science and technology, we (especially those of us in the most advanced economies) are living at a time of unprecedented power to prevent and treat human diseases. This is so because fundamental science is providing a deeper understanding of the biological mechanisms by which diseases develop and because novel technologies are allowing us to interfere with those mechanisms.

In this new book, William Pao – an accomplished physician, respected basic scientist and leader in the pharmaceutical industry – tells us eight fascinating stories about modern therapies that range widely from structurally designed chemicals to edited genes to enhanced antibodies. These stories reveal how new medical treatments arise from advances in many branches of science and medicine, including biochemistry, genetics, chemistry, structural biology, pharmacology, computational sciences and clinical trials.

Happily, the stories are told in language that is accessible, without oversimplifying the science that has fuelled such remarkable progress.

While most of these stories do not yet end with the elimination or the cure of a disease, they are inspiring as adventures of the imagination, illustrations of human ingenuity, and demonstrations of how science can improve lives. In that sense, they have the potential to attract talented youth into the relevant fields of medical science, just as another book about discoveries of the principles of infectious diseases, *Microbe Hunters* by Paul de Kruif, was said to have done almost a century ago.

Comparison of these two books is an enlightening way to recognise the profound changes in medical science that have occurred during the past hundred years. De Kruif, an infectious disease expert who once worked at the Rockefeller Institute, wrote engaging, heroic tales of individuals – incidentally, but not surprisingly, all males – who discovered some of the first known microbes, mostly in the late nineteenth century; identified several of them as agents of terrible diseases (including tuberculosis, syphilis, malaria and yellow fever); and began to learn how such diseases might be controlled by our immune systems, by vaccines, or by chemicals found serendipitously to be toxic to the pathogens.

In the stories told here, only a few of the diseases are caused by invading microbes. The others are cancers or inherited disorders, affecting a variety of tissues and functions. And the heroes are not simply individuals. They are teams, composed of various kinds of people: scientists of several stripes, working in academia or industry; physicians caring for patients or conducting clinical trials; and others who support research as advocates, as funders, and as employees and leaders of public and private research facilities.

The teams may work synchronously to solve a problem. Or they may confront a series of difficulties sequentially, passing the baton from one set of experts to another. This intricate relay can happen in several stages: when a new disease, or new form of a known disease, is first identified and needs to be better understood; when the mechanism of a disease has been elucidated and its vulnerabilities need to be sought; when a potential target for therapeutic intervention is proposed and requires validation; or when a possible therapy needs to be manipulated chemically, evaluated in animal models, or subjected to definitive testing in human patients. The elaborate teamwork required to carry out these several complex steps may seem inherently different from the solitary actions taken by de Kruif's heroic adventurers. But the rewards for patients and the public prove to be at least as great, and the stories at least as interesting and informative, as those told a century ago.

When readers of this book reach the middle of Chapter 2, they will learn that I know more about William Pao from direct experience than from hearsay and from reading this book. We worked together during the first stages of an exciting period of research on human lung cancer, when our findings and results from other labs drove rapid changes in the treatment of this common and frequently fatal disease. So I am well positioned to confirm that his skills as an investigator and clinician confer authenticity to his voice as a narrator of these tales – especially when supplemented by his extensive, more recent experiences as a leader in the pharmaceutical industry. Reader, you are in good hands – about to be told some remarkable stories by someone on the frontiers of medical science.

Harold Varmus, MD,
New York City, January 2024

INTRODUCTION

THE BRAZILIAN PIT VIPER IS A particularly lethal predator. It has long posed an occupational hazard for farmers in south-eastern Brazil who work in its natural habitat. Over the centuries, many witnessed the terrifying effects of its venom: one bite can cause a person to collapse on the spot. The venom is so potent that indigenous tribes used it to coat arrowheads to disable prey.

Back in the 1940s, the pit viper drew the interest of a Brazilian pharmacologist named Maurício Rocha e Silva. While much of the world was engulfed in war, Rocha e Silva was researching circulatory shock at the Biological Institute in São Paulo. His team sought to understand the toxicology of snake bites, unravelling how venom acted on the human body.

In 1948, they identified a previously unknown peptide that became elevated in blood plasma after animals were dosed with pit viper venom. (A peptide is a short chain of amino acids, the building blocks of proteins.) The mysterious molecule caused blood vessels to dilate: once a victim had been bitten, blood pressure dropped, sometimes catastrophically. Without sufficient pressure to force blood around the body, vital muscle, nerve and brain cells were starved of oxygen.

Rocha e Silva and his team named this strange, havoc-wreaking peptide *bradykinin*.[1]

Sérgio Ferreira was just fourteen at the time. Growing up in the state of São Paulo, he later applied to medical school with the ambition of becoming a psychiatrist. He changed his mind when confronted with the reality of his dream job: 'Public psychiatry care in Brazil was rather poor, so I decided to become a scientist.'[2] He joined Rocha e Silva's lab and was put to work investigating pit viper venom and bradykinin.

In 1964, for his PhD project, Ferreira showed that pit viper venom contains a substance, *bradykinin-potentiating factor* (BPF), that makes bradykinin much more active.[3] Ferreira's discovery further confirmed that pit vipers are lethal because they subvert the vital molecular system that regulates our blood pressure.

But the true medical breakthrough came when Ferreira moved to London to join the lab of another distinguished pharmacologist, John Vane. Ferreira took with him a vial of the pit viper BPF.

Vane, the grandson of Russian immigrants, was a self-described 'experimentalist'. 'At the age of twelve,' he wrote, 'my parents gave me a chemistry set for Christmas and experimentation soon became a consuming passion in my life.' His first experiments made use of a Bunsen burner fed from his mother's gas stove, but 'a minor explosion involving hydrogen sulphide' (a toxic and corrosive gas) stained the kitchen's newly painted walls, and the precocious young scientist was banished to a shed.[4]

Vane was interested in high blood pressure, a primary driver of mortality worldwide. Hypertension is a major cause of strokes, heart attacks, and heart and kidney failure. At that time, millions of people were at risk of premature death because they had no reliable way to control their blood pressure. Our bodies must be able to raise or lower our blood pressure, depending on our physical activities, and we have a complex system of physiological and backup controls to keep our blood pressure at an appropriate level. Vane and his team were busy identifying some of the key components of that control system when Ferreira arrived. One key component is a blood pressure-raising peptide called *angiotensin II*. To make this peptide, we use an enzyme called *angiotensin-converting enzyme* (ACE).

'Sérgio Ferreira brought this ... brown goo,' recalled Mick Bakhle when interviewed in 2016, laughing at the memory. Bakhle was another member of Vane's team, assigned to investigate ACE at the time. On hearing of the effect of pit viper venom on blood pressure, Vane asked Bakhle in 1970 to test Ferreira's BPF on ACE. 'Brown goos are not very nice to work with, but we did have a look at it. And much to our surprise ...'[5]

Ferreira's brown goo turned out to inhibit ACE. It was an extraordinary discovery: an extract from a South American snake venom was shown to knock out the key enzyme that produced a critical blood pressure-raising peptide. Without ACE, there's no angiotensin II; without angiotensin II, there might be no high blood pressure. ACE was also found to be the enzyme that inactivates bradykinin, meaning that without ACE, bradykinin would also remain around to lower blood pressure. This latter finding completed the pit viper puzzle.[6]

John Vane understood the potential medical significance of the ACE-inhibitor in pit viper venom immediately. But he also knew that Ferreira's peptide, potent as it was, would make a lousy blood pressure drug. High blood pressure is a chronic condition that needs to be treated regularly over a long period of time, perhaps for an entire lifetime. It's very hard to persuade most people to take a medicine repeatedly any other way than orally. But BPF could not be taken orally – the delicate molecule would be broken down in the stomach long before it reached the bloodstream. It could only be administered the way pit vipers do it – by injection.

What the world needed was an ACE-inhibitor that could withstand the human digestive system and be absorbed whole from the gut into the blood: a blood pressure treatment in a pill.

Such a pharmacological pearl was beyond the capability of an academic lab. But Vane had a side gig as a consultant to the American pharmaceutical company Squibb, and he suggested

ACE-inhibitors as a potential research avenue. Two Squibb chemists, Dave Cushman and Miguel Ondetti, took up Vane's lead. They mapped the molecular structure of Ferreira's pit viper peptide and set about designing a more robust chemical cousin. It took them many years and multiple setbacks, but eventually they succeeded. The drug they came up with was named *captopril*. It was approved by the US Food and Drug Administration in 1981 and was the first reliably safe treatment for high blood pressure.

At first glance, captopril seems a modest kind of molecule made of commonplace components: nine carbon atoms, fifteen hydrogen, three oxygen, one sulphur and one nitrogen – that's all. Yet, captopril was perhaps one of the most important innovations of the twentieth century in any field of scientific discovery. Aeroplanes opened up the world; semiconductors and the internet ushered in the digital age; but captopril and the other ACE inhibitors it inspired have given millions of us longer lives.

I'm an oncologist – a cancer doctor. I treated very sick patients for fourteen years. During that time, I prolonged the lives of many patients, but I also saw many others die or suffer protracted pain and incapacity. Each patient provided me with inspiration to improve on the status quo, but the biggest inspiration for my lifelong work has been my dad, who died prematurely from colon cancer in 1981.

Born in China in 1922, my dad set sail in May 1948 on the American President Lines' U.S.S. *General Meigs* to the United States. He had no relatives in this new country, but with a degree in hand from the National Medical College of Shanghai,* he completed residency training in Ogden, Utah, on a stipend of $100 per month. After further training in Baltimore, Maryland, he got a job at Chestnut Lodge, a private psychiatric hospital in Rockville, Maryland. He eventually became Director of Psychotherapy.

* The National Medical College of Shanghai later became the Shanghai Medical University; in 2001 it was integrated into Fudan University.

Some of my best memories of him were from playing cards or Scrabble, hearing him sing Broadway tunes, and seeing him perform in Chinese operas. I also fondly recall that every four weeks, he and I would go and get our hair cut and have lunch together.

The first time Dad was admitted to the hospital, in 1979, I was eleven years old. I don't remember being told that he had been diagnosed with cancer. I think he had found blood in his stool. Looking back, I suppose he had stage III colon cancer which was resected surgically.

Afterwards, he was put on what must have been 5-fluorouracil (5-FU), a chemotherapy drug that made his hair fall out and left him nauseated and vomiting. That was the end of our shared haircuts and lunches for a while. Following chemo, though, he bounced back and life seemed to return to normal.

About two years later, Dad returned to the hospital for exploratory surgery. The cancer had reappeared, and the doctors wanted to determine how far it had spread. (Radiographic imaging was not very advanced at that time.) He died on the operating table. The surgeon had found widespread liver metastases. After biopsying a lesion, he couldn't stop the bleeding. Maybe this was a small mercy: my dad evaded the slow agony of cancer's ravages.

For my family, the loss of our father was a cataclysmic shock. But our story is by no means unique. Cancer still claims a horrendous number of victims every day, too many of them young, too many of them leaving behind grieving family and friends. Around ten million people die of cancer worldwide each year – almost one in six of all deaths.[7]

My dad and my mom, also a physician, had always expected that my older sister and I would go into medicine. Our father's death only reinforced that trajectory for both of us. After Dad died, I vowed that I would dedicate my life to making a difference for patients like him.

About twenty years later (after college, medical school, internship, residency, fellowship and post-doctoral training), I became a medical oncologist and translational science researcher – a cancer physician-scientist. I treated patients while at the same time running an academic translational research* lab to figure out at the molecular level why cancers grow, and ultimately to find ways to kill them.

Throughout that time, I collaborated with pharmaceutical companies to develop new medicines, including an important new lung cancer drug that is now prolonging the lives of many patients (see Chapter 2). In 2014, I left the world of academic medicine and transitioned to leadership roles in research and development in the pharmaceutical industry. I wanted to help create new therapeutic options directly, to make a bigger impact for millions of patients worldwide.

I've learned over the years that a great deal of creativity goes into making new medicines, most of it witnessed and appreciated by only a small handful of people. Across the wider pharmaceutical industry, amongst biotech firms and government labs, and throughout academia, thousands of scientists and experts are striving daily to create molecules that will save lives and help people feel or function better. They are trailblazing new areas of biology, inventing new tools and technologies to hit drug targets, and advancing our understanding of diseases. This is a community fuelled by innovation, a world of drug hunters charting new scientific territory every day of the year. Yet almost all of it remains invisible to the public.

To me, innovation is coming up with something that hasn't been done before, showing that it's possible, and having the courage to convince others that it's worth doing. It's something that many of us seek to do in all walks of life. We inevitably will look for ways to do things better – *to add value through invention.*

As a cancer physician, research scientist and drug developer, I have been witness to a vast field of innovation which is, at best, impenetrable to most people; at worst, it goes unnoticed. The men

* Translational medicine seeks to convert laboratory discoveries into practical medical applications and to discover the molecular mechanisms underlying clinical phenomena observed in patients.

and women working on the next generation of medicines are undertaking some of the most extraordinary innovation ever seen. I want to lift the veil on some of that innovation and share stories from the front line of drug discovery.

It is not a straightforward ambition. The product of a Silicon Valley innovation story is often a super-branded and seemingly ubiquitous device or app that is intuitive to use. By contrast, the product of a pharmacological innovation story is a molecule too tiny to see, with an alien name and a mode of action that will often seem incomprehensible even to people with science degrees. That molecule may save thousands of lives, but if it's a nightmare to pronounce, who's going to talk about it?

And let's face it: most people prefer not to take medicines if they can possibly help it; they learn about these drugs unwillingly – when they or their loved ones are struck by a disease. Furthermore, since everyone's health is different, only a few medicines are widely known or taken.

As we glimpsed from the story of captopril, there is seldom just one hero or one team responsible for a new medicine. Hundreds of scientists, working across multiple organisations and decades, usually play a part in the innovation. Even the doctor who prescribes your medicine most likely will not know who invented it. That makes the stories of drug discovery both abstract and opaque compared to other examples of innovation.

But the toughest challenge may be the science itself. Rocket science is famously difficult, but the fundamental problems of space exploration are straightforward to communicate: how to ensure enough oxygen for the astronauts; how to withstand the cold in space, or the heat of re-entry; how to move in zero gravity. When it comes to treating cancer, however, the storyteller needs to start by explaining what cancer is at the cellular level, how it's driven by molecular signalling pathways consisting of enzymes encoded by oncogenes and, by the way, what are all those things?

Nevertheless, I'm going to try to share some of the absolute awe I feel for the astounding innovation taking place in drug discovery labs around the world. These stories open a door to a more detailed and nuanced appreciation of what it takes to create something new and valuable that we trust to act on – and *inside* – our bodies. Medicine is one of the first technologies we encounter as children; we think it is normal to take a tablet to make us better. Yet, what a remarkable thing that is! As one of the scientists we shall meet later exclaimed, '*Wow, oh my God! A serious disease can just disappear if a chemist builds the right molecule.*'

To make a new drug we must decode nature – the biological secrets of life that have evolved over millions of years. We must identify and characterise a disease, understand scientifically why it happens, and then find a way to alter its course by giving a patient a particular molecule that will impact the disease without incurring significant side effects. The whole process is a triumph of human ingenuity, perseverance, collaboration and resilience. Reading these stories should fill you with hope for the potential of humankind to make the world a better place.

The science is challenging, it's true, but it's not impenetrable. In fact, I would argue it is fascinating, sometimes thrilling, occasionally revelatory. Once you've understood it, you'll have a far better idea of how your own body works. You won't need any scientific education to follow these stories, just a willingness to discover.

My hope is this book will serve as a bright, welcoming beacon for young people considering a career in the life sciences. Our mindsets, and therefore all the big decisions we make, are shaped by the stories we hear, and there are just too few stories about those who invent medicines.

If we want to encourage the next generation to join in the battle against cancer, to take on dementia, to be ready to respond to the next pandemic, we need to be unearthing and sharing the stories of exemplary scientific innovation by these unsung heroes.

So come with me on a deep dive into eight drug discovery adventures. We'll find out why paracetamol, one of the world's most popular drugs, was left unused for decades after its discovery. The extraordinary tenacity of a sick child's mother will illustrate how innovation can depend on the determination and drive of a few individuals. A cunningly modified part of the immune system will be the breakthrough that frees people with haemophilia from the tyranny of near-daily injections. A type of blood we normally see only inside the womb will be resurrected in adults to treat sickle cell disease. A passion for the geometry of cones will lead to the discovery of a new HIV treatment. A worldwide pandemic will spur the discovery of an antiviral in record time. And the lives of patients with cancer will be prolonged by new therapies against specific genetic targets.

Along the way, we'll discover where the big ideas come from, how the best scientists overcome obstacle after obstacle, how diverse teams work together in pursuit of a common goal, how innovators make the most of serendipity, how breakthroughs depend on a foundation of deep, seemingly unconnected knowledge, how even the smartest scientists depend on trial and error to make progress, and how personal curiosity and commitment keep people going.

Drug discovery shows that seemingly impossible breakthroughs can be achieved, given time, dedication, skill, collaboration and a dash of luck. I hope the dogged determination and wondrous creativity shown by these drug developers inspires you to innovate in the field of your choice, and even change people's lives, as they have.

GETTING STARTED

MEDICAL SCIENCE IS A COMPLEX TOPIC. To ease that complexity, here's a quick briefing on some key biological concepts and an outline of the drug development process. These introductory notes are here if you find yourself getting lost in amino acid chains or pre-clinical toxicology reviews further down the road. You can also consult the glossary at the back.

A Crash Course in Biology

To understand how drugs work, we must go deep into the science of cells, biochemical systems and ultimately molecules – the actual targets of the drugs.

We can start simply. Our bodies are made up of the things we eat and drink: water, sugars, fats, proteins and minerals. For drug hunters, it is the *protein* that is of the greatest interest. Proteins come in a dazzling array of forms. They make most of the important stuff happen in the body – catalysing, signalling, metabolising – and so we generally seek to *activate* or *inhibit* proteins. That's what most medicines do: they encourage particular proteins to do something, or they prevent proteins from doing something.

One category of key proteins for the drug hunter is enzymes. An *enzyme* is a protein that catalyses some kind of change. That change happens when the enzyme *binds* to one or more other molecules. That means every enzyme has at least one *binding site*, a physical docking bay where small natural molecules – or drugs – can bind.

Proteins are built from blueprints encoded in *genes*. Our genes contain manufacturing instructions written in strands of *DNA*. Each gene encodes a specific protein. DNA is first 'transcribed' into messenger RNA (mRNA), and then mRNA is 'translated' into protein.

Confusingly, the protein and gene often have the same name. You can tell the difference in print because the gene is italicised. So, *EGFR* is the gene for EGFR protein. I'll try to make it clear when I'm talking about a gene or a protein.

Proteins are made up of chains of *amino acids*.* These are very simple molecules, usually consisting of a few atoms of carbon, hydrogen, nitrogen and oxygen. Different proteins have different sequences of amino acids, and these sequences determine the protein's unique structure, which in turn determines its biochemical properties – how it behaves in the body. When individual amino acids come together in a chain, they undergo a slight chemical change; the resulting links in the chain are known as *amino acid residues*.

Biological molecules and systems, by their nature, are often complex and unpredictable. Living things evolved over hundreds of millions of years, and we just don't understand everything about them yet. So, when we intervene in a biological system with a new potential drug, we may encounter surprises. Drug discovery might be summed up as designing molecules to manipulate biology – but biology is not always so easily manipulated.

* Humans have twenty different types of amino acid.

How Medicines Are Made

Drug discovery and development is the process of inventing and testing a new medicine to treat a particular condition or disease. At the heart of a medicine is usually a single *active ingredient* that affects a specific target and does the work of treating cancer, fighting infection, and so on. There may be other components that help the medicine get to the right part of the body and do its job effectively. The active ingredient is a unique *molecule* – a collection of atoms arranged in a particular way. This may be a *small molecule* (generally under 1,000 atomic mass units), a chemical typically created and synthesised by chemists that can often be taken orally in tablet form. Or it may be a *large molecule* (generally over 150,000 atomic mass units), a *biologic* such as an antibody, created by protein engineers and produced in living cells. Due to their large molecular size and other intrinsic properties, biologics are usually injected. The type of molecule selected for a particular medicine depends upon the specific target of interest.

When we want to make a new medicine for a disease for which there is still an unmet medical need, we start by trying to understand how the disease works in the body. If it's an infectious disease, how does the pathogen get into the body, how does it cause harm, how does it replicate and spread? If it's cancer, how does it start and what makes it grow? If it's a genetic disease, what's the genetic abnormality causing it, and what exactly is it doing in our cells?

'Basic science' activities address questions like these. They're 'basic' because we're trying to identify the foundational biochemical mechanisms behind biological phenomena. Much of basic science takes place in universities and research institutions and is funded by governments or non-profits. However, it also occurs in biotech and pharmaceutical companies. The answers to many basic questions have no immediate therapeutic implications, and it may take decades for a biological insight to lead to a drug.

Once we have some understanding of the disease, we look for a way to address it. For viral infections, we may want to interfere with viral replication. For cancer, we may want to stop tumour cell proliferation. For genetic abnormalities, we may want to correct the faulty mechanism. To do the research in a safe manner, we work outside the human body: we try to reproduce a critical element of the disease in a test tube, petri dish or animal model. By design, our models are highly simplified versions of the human disease; we just can't mimic all aspects of a disease in the laboratory yet.

If we don't have a promising lead like pit viper venom, we may use trial and error on a grand scale to seek a molecule that might disrupt or correct the disease. Trial and error has a noble history in pharmacology: early drug hunters could do little more than try out different chemicals on patients to see if any had a positive effect. Now, we can test millions of different molecules against an isolated representation of the disease in a *high throughput screen* and look for some sort of desired biological response.

If we find a molecule that seems to do what we want, we'll look through a whole family of related molecules to see if we can identify one that gives a better, stronger response. Eventually, we identify a *lead candidate molecule*. We then might need to modify that molecule, adding a couple of atoms here, removing a group of atoms there, to make it more potent, less toxic, more soluble, or better able to withstand the body's digestive processes.

Throughout this process of improving and optimising the molecule, we use *assays*, or tests, to measure how we're doing. The assays might be *in vitro* – testing the molecule on isolated proteins or cells in a test tube or petri dish – or they might be *in vivo* – testing in a live animal such as a mouse, often genetically modified to have a simplified form of a disease.

Gradually, incrementally, this process of designing, making, testing and refining molecules will, we hope, eventually yield an experimental drug (*a development candidate*) that merits testing in humans. The whole process leading up to this point is the *Discovery* phase. Some of it may take place in a university or other research institution, but most drug discovery takes place in pharma or biotech companies.

Next comes the *Development* phase, where we finally test the experimental molecule in humans. This usually takes many years and can cost hundreds of millions of dollars to complete. To understand why, it's helpful to remember how things used to be.

Before the 1900s, rules and regulations around medicines were not yet established. Snake oil salesmen would roll into town and hold up a bottle of some bitter-tasting or sweetly alcoholic fluid and declare it to be an astounding cure for headaches, or scurvy, or cancer, or impotence, or all of the above. You would pay your money and swallow the substance, and you would conclude it must be a real medicine because it had a violent effect on you – making you shiver or vomit or sweat. It might be poisonous, but you'd have no way of knowing that. You wouldn't really know if it worked either, because if you did get better there was every chance your recovery happened in spite of the 'medicine'. If it really did work, and you wanted more, you would have no idea if the next bottle you purchased contained the same dose, or even the same active ingredient.

Regulation and responsible clinical testing have transformed the entire drug development process. Exhaustive testing of the molecule is now required for a regulator such as the US Food and Drug Administration (FDA) or the UK Medicines and Healthcare products Regulatory Agency (MHRA) to approve a new medicine.

For example, before we can give the 'new molecular entity' to a single human, we are required by law to put it through a *preclinical toxicology review*. If we're developing a small molecule, we have to test it for safety in two species of animal.

Once we pass toxicology testing, we administer low, gradually increasing doses to a small number of (usually healthy) volunteers, looking to see if the molecule hits its target and/or causes worrying side effects. We take blood samples to understand how the molecule behaves and moves around the human body. We start with single doses. If side effects are acceptable, we go on to administer multiple doses. Traditionally, this initial study in humans is called *Phase I*.

Next, we test the molecule for efficacy in patients with the disease, closely monitoring for adverse side effects. We trial increasing doses of the compound and use the safety and efficacy data from these different doses to select the best clinical dose. This is *Phase II*.

Most experimental drugs fail at one of these *early development* phases: they are found to be either ineffective or unsafe. Perhaps our laboratory models were not realistic enough. Perhaps human subjects suffered a side effect which could not have been predicted from our pre-clinical toxicology experiments. If, however, the clinical *investigators* (physicians independent of the companies), drug developers, regulators and company decision-makers are satisfied that the molecule is both safe and potentially an improvement on the current standard of care, the molecule enters the final *late development* stage, *Phase III*.

Now the molecule is administered to larger numbers of patients, often in multiple countries, usually in double-blind randomised controlled trials that compare the effectiveness of the compound against a placebo or an existing treatment. Again, safety is closely monitored, and any side effects or *adverse events* are painstakingly noted and reported to regulators. This *pivotal* phase is the most costly one, because hundreds or even thousands of patients are treated and monitored over a period of months or years. Finally, if everything goes well, the compiled patient response and safety data, along with drug manufacturing data and other relevant materials, are all submitted to regulatory agencies.

Once regulators have examined all data and found clear evidence of safety and efficacy along with reliable, consistent manufacturing processes, they approve the new medicine for sale in specified markets.

Due to the cost and complexity of the development process, large Phase III studies are mostly carried out by big pharmaceutical companies. Few entities have the resources or the organisation to run them.

Animal Testing

From time to time, I will refer to results from animal studies where they impact the discovery and development process.

Since the advent of drug regulatory agencies, laws in the United States, European Union, the United Kingdom, Japan and most other jurisdictions have stipulated that every new molecule has to be tested in animals before it is given to humans.*

We use animal testing primarily to ensure that molecules are as safe as possible before they are administered to people. We also look for proof of efficacy in animals. For example, mouse models of human cancers are used to determine if tumours shrink after treatment with an experimental compound.

Most tests are carried out in mice and rats. All animals are cared for as humanely as possible. All animal testing is carried out in strict accordance with internationally agreed rules and overseen closely by various international, national and local organisations.†

Big Pharma

This is not a book about pharmaceutical companies or the pharma industry in general. It is about the innovation process behind drug discovery and development – wherever that happens – and the stories of the science and the scientists behind different molecules. Nevertheless, each of the eight drugs we'll look at was developed in part by at least one pharma company.

* In 2023, the United States eliminated that requirement, and pharma companies are now considering alternative options.

† Across the Life Sciences, we are all committed to the 'Three Rs': Replacement, Reduction and Refinement. Replacement means searching for methods to replace the use of animals. Reduction means pursuing strategies to answer our research questions with fewer animals. Refinement means modifying husbandry and experimental procedures to minimise pain and distress, and to enhance the welfare of a research animal from the time it is born until its death.

While I was writing this book, I was employed by two different pharma companies. *Breakthrough* is not sponsored by, or written in support of, either company. The extent of their involvement is that both have allowed me to interview their scientists and feature one of their new drugs.

All eight drugs in this book are remarkable in some way. However, I won't extol their various attributes for two reasons. Firstly, this is a book about discovery and innovation, not a marketing pamphlet. Secondly, in many jurisdictions it is illegal to promote prescription medicines directly to the public. Please understand this unique challenge of writing about pharmaceuticals. Consult your doctor or the relevant health authorities if you would like to know if these medicines are appropriate for you or a particular patient. For specific information (including efficacy and safety data) on these medicines, please see the relevant prescribing information at the regulator website in your country.

On the Naming of Drugs

There's no getting around it: drug names are not sexy. We complicate matters by giving drugs two different names – a generic name and a brand name. This is necessary because the same drug might be sold by multiple companies, with each brand exclusive to one company. So the painkiller *ibuprofen*, which was invented by a team of scientists at Boots UK Ltd, is marketed as *Advil* by Haleon plc and as *Nurofen* by Reckitt Benckiser Group plc. Another reason to distinguish between the generic name and the brand name is that some medicines include multiple molecules. For example, *Paxlovid*, Pfizer's oral treatment for Covid-19, is a co-administration of two drugs, *nirmatrelvir* and *ritonavir*.

You can tell the difference between generic names and brand names easily in written text: brand names are capitalised. So *alpelisib* is the generic name, *Piqray* is the brand name. To avoid confusion or potential charges of unethical marketing, I will use the generic name wherever possible.* Technically, the generic name is the *international nonproprietary name* (INN), allocated under a convention governed by the World Health Organization.

You may wonder why generic names have to be *so* alien and difficult to remember. If you look more closely, though, you'll find that there is a logic to them. For example, all *antiviral* drugs end in *-vir*. Those antivirals that target a virus's 3CL protease (an enzyme) end in *-trelvir* (see Chapter 6), while those that target a viral capsid end in *-capavir* (see Chapter 8).

The Credits

These stories are bursting with names. No one is expecting you to keep track of them. One of the key themes we'll be exploring is the collaborative nature of innovation: hundreds, even thousands, of people played a part in the discovery and development of each of these drugs.

The main actors, with oversimplified job descriptions, include:

Biochemists, who study how molecules operate and test them for efficacy in test tubes;
Biologists, who study how diseases arise and test molecules in living cells or animals;
Chemists, who design and make chemical molecules (small molecules);
Executives, who make key strategic, talent and funding decisions to enable all of the work;
Geneticists, who study the genes and genetic mutations that cause disease;
Pharmacologists, who assess drug metabolism, drug stability and pharmacokinetics;

* Where interviewees referred to a drug by its brand name, I have substituted the generic name, even in direct quotes.

Physicians, who design and/or conduct clinical trials for the new molecules in humans;

Protein engineers, who design and make protein molecules (large molecules);

Statisticians, who help correctly design experiments, verify conclusions and properly interpret results;

and **Toxicologists**, who test and help design molecules for safety.

There are so many more experts involved, but space does not permit me to list them all.

In each story, I have focused on just a handful of people. The majority of contributors to these drugs have not been named here for reasons of space. To them I apologise and ask for their forgiveness and understanding.

Without further ado, let's explore their breakthroughs.

CHAPTER 1

The World's Most Common Rare Disease

Condition: Spinal muscular atrophy
Gene: *SMN2*
Innovation: Risdiplam [RIZ-di-plam]
Companies: PTC Therapeutics and Roche

LOREN ENG WAS TRYING TO ENJOY the dinner party. It wasn't as easy as it used to be. Loren was seven months pregnant with her second child. Like any parent, she was worn out by the sleepless nights and endless little anxieties that came with caring for her firstborn, Arya. And she had an extra worry: at seventeen months, Arya still wasn't walking right. Her gait was stick-straight, like a toy soldier's. Every few steps she would stagger, and occasionally even collapse.

Well-meaning friends and physicians told her not to worry. Lots of kids lag behind on something – speech, walking, social skills. She'd catch up soon enough, Loren was assured. Enjoy being a parent! Arya was a beautiful and happy little girl. What did it matter if she took a little longer to get the hang of walking?

The New York dinner party was hosted by an old college friend and her mom, who happened to be a paediatrician. After they'd cleared away the dishes, Loren couldn't help but ask:

'This walking thing . . . it's normal for some kids to develop a little late, right?' She glanced over to the couch where Arya was happily flipping through a picture book.

Her friend's mom looked at her kindly. 'What does your doctor say?'

'That I'm worrying too much,' admitted Loren. She'd always been careful, always taken every last precaution. Before Arya was born, she'd undergone all the available tests for genetic disorders. 'Arya had perfect [Apgar] scores at birth,' she would later recall.

Yet here was her daughter at seventeen months, stumbling like she was intoxicated.

The paediatrician set aside her glass of wine and got down on her hands and knees. She introduced herself to Arya and offered her a spoon to hold. In a gentle, easy-going manner, she encouraged the little girl to pick up objects, to squeeze her finger, to crawl, to walk. There and then, while the other guests chatted over coffees and herbal teas, she examined Arya.

Finally, she straightened up. 'There's something very wrong,' she said with a panicked look. 'You need to get her to a neurologist right away.'

It was August 2001, just a few weeks before the city – and the world – would be shaken by the attack on the World Trade Center. Loren had to wait through all of that horror and turmoil for the test results that would reveal the awful truth.

Arya was diagnosed with spinal muscular atrophy (SMA). The disease would eat away at her strength, bit by bit, until she could no longer move, perhaps no longer breathe, without mechanical help. There was no cure. SMA would put her in a wheelchair, would paralyse her muscles and, ultimately, might very well kill her.

SMA was first identified by Guido Werdnig and Johann Hoffmann in the last decade of the nineteenth century. The two neurologists, one Austrian and one German, independently described a previously unknown condition afflicting infants. They saw babies who should have been growing stronger with age start to get *weaker*. Their tiny hands would tremble, they would lose the ability to grip their mothers' fingers, they would struggle to eat and be unable to sit up or even raise their head. The result was a 'floppy baby'. In the most severe cases, life expectancy was less than two years.

Werdnig and Hoffmann recognised that SMA was a genetic disease – caused by faulty genes inherited from our ancestors. And they noted one other important characteristic: it sometimes manifested in children whose parents were both completely healthy. This implied the genetic trait was *recessive*.

We each have two copies of most genes. If a genetic condition is recessive, both copies of the relevant gene must be faulty for the condition to occur. Anyone with a single copy of the faulty gene is an unaffected *carrier*. When two healthy parents are both carriers, each of their children has a twenty-five per cent chance of inheriting both copies of the faulty gene and so developing the condition.

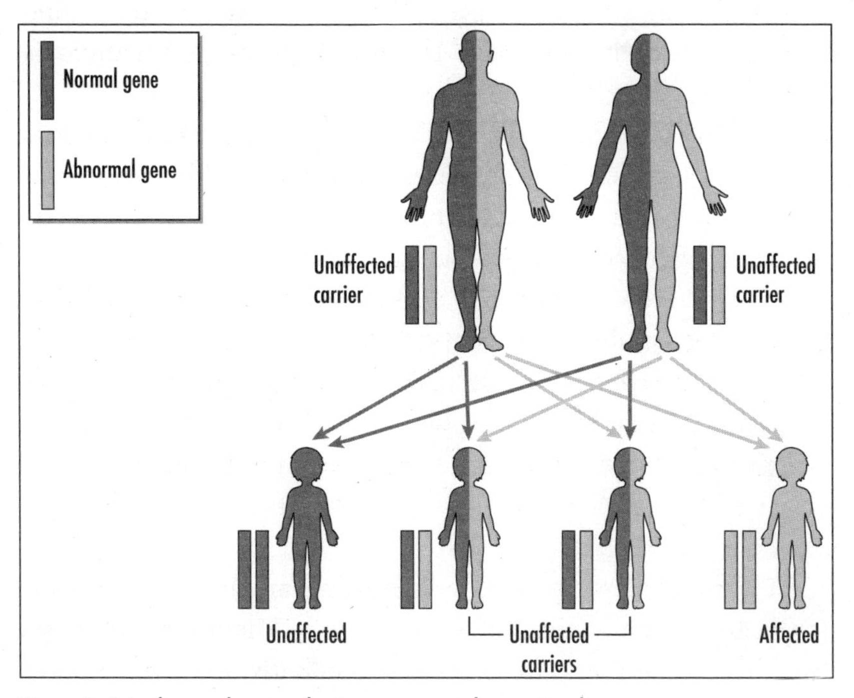

Figure 1. Spinal muscular atrophy is an autosomal recessive disease.

These two characteristics of SMA – genetic and recessive – represent its two opposing faces. SMA's recessive nature is a boon to humanity: although as many as one in forty of us are carrying the faulty gene that causes SMA, most of us are untouched by it. Only about one in eight thousand infants worldwide is diagnosed with the disease.[1] But because SMA is genetic, for those who are unlucky enough to inherit both copies of the faulty gene, there has until recently been absolutely no hope of relief. For most of human history, what is written in the genes has been impossible to change.

Loren Eng understood exactly what this meant for her daughter when Arya's genetic analysis revealed the grim news. The two copies of the gene that should have been encoding a protein vital to building and sustaining Arya's motor neurons were both faulty. Without a regular supply of the protein, her motor neurons – the long nerve cells that controlled her muscles – were deteriorating. As her motor neurons degenerated, the muscles they should have been activating were withering away through lack of use.

There was nothing that could be done for her. No cure. No therapy. All the doctors could advise was to take Arya home and try to give her the best life possible for her remaining years, comforting her as she became ever more paralysed.

Thankfully, Arya's younger brother, born a few days after her diagnosis, escaped that genetic short straw. That was one small mercy for Loren as she faced up to the reality of her daughter's prognosis. Arya did not have the most severe form of the disease, so she would live past her second birthday. But what kind of life could she hope for? Would she be able to attend school? Would she ever play sports? Would she get to go to prom? How on earth could Loren explain to her daughter the inexorable physical deterioration she would suffer?

One thing distinguished Loren and Arya's father, Dinakar Singh, from other parents of SMA children. Having arrived in the United States as immigrants, they had built careers in finance and business strategy, providing them with the means to fund and direct scientific research themselves. Within just a few weeks, Loren, her family and friends had vowed to dedicate themselves to finding a treatment for Arya and the many others like her.

A few years earlier, that ambition would have seemed a fool's errand. Very little was known about SMA, except that it was a genetic disease and therefore essentially incurable. But just six years before Arya was diagnosed, science had achieved an important breakthrough: the gene for SMA had been identified.

The last decade of the second millennium began a golden age of discovery in the science of genetics. The Human Genome Project was launched in 1990, and a 'rough draft' of the entire human genome would be completed by 2003. University and commercial laboratories all over the world were racing to discover our species' coded secrets. One of these labs was run by Judith Melki.

Before moving into research, Melki had worked in a hospital as a junior doctor, and she had been struck by the tragic fate of the SMA patients on her wards. 'We had nothing to offer them,' she says. 'Very young children were condemned to die. Well, as a young physician you want to change this.'

Melki took a year out from clinical practice to research neuromuscular disorders. She wanted above all to identify the gene responsible for SMA. At the French National Institute of Health and Medical Research, she used a technique called linkage analysis to work out which fragments of chromosomes were most commonly shared in families where two or three children had SMA. Her team localised the gene to Chromosome 5, band q13.* This part of the chromosome turned out to be highly complicated, with lots of duplicated genes. But only one was critical to SMA.

Melki and colleagues were able to show that in ninety-five per cent of SMA patients, this gene was either damaged or missing. The gene encoded a previously unknown protein which they

* Humans have twenty-three pairs of chromosomes, numbered one through twenty-two. Half come from the mother and half from the father. The last two chromosomes are the sex chromosomes, X and Y.

named, with macabre clarity, 'Survival of Motor Neurons' (SMN). The gene itself was labelled *SMN1*.

Melki's team discovered a second, almost identical gene that also produces SMN protein. Unfortunately, this second gene does a really bad job of it. Where the *SMN1* gene pumps out SMN protein by the bucketful, all the *SMN2* gene can manage is a thin trickle. For most of us, the *SMN2* gene is irrelevant.

On the other hand, if your *SMN1* gene is damaged or missing, *SMN2* suddenly looks like a godsend. That thin trickle of SMN protein it produces is the only thing keeping your motor neurons alive. *SMN2* is the reason why SMA patients survive at all. Some people have multiple copies of the *SMN2* gene, and the total amount of SMN protein they can produce is therefore a little higher, explaining the different levels of severity of the disease.

Melki's discovery of the *SMN2* gene, feeble as it is, offered the first hint of a possible route to treat SMA. If children like Arya didn't have a functioning *SMN1* gene – if they were entirely dependent on the thin trickle of SMN protein stemming from the *SMN2* gene – what if we could find a way to open up the spigot and turn that trickle into a reliable flow? What if we could make the human body do something it had never done before?

Imagine a production line churning out automobiles. Every car has a perfect left headlight, but the right headlight is always dysfunctional, emitting just a glimmer of light. The left headlight produces more than enough light to drive safely at night, so the deficient right headlight is ignored. But now imagine something goes wrong on the production line and the next car produced has no left headlight at all. Suddenly, the right headlight becomes interesting. As things stand, it doesn't emit enough light to drive safely at night, but perhaps we could change that. What exactly *is* wrong with it? It's emitting a bit of light, so the power supply and bulb must be in working order. Wouldn't it be worth taking a look inside to see if it can be fixed?

Wouldn't it be worth taking a look at that dysfunctional *SMN2* gene?

* * *

Bleak as the SMA medical landscape was in 2002, parents, physicians and researchers could see a glimmer of hope in that second *SMN* gene. 'It was a one in a million stroke of good luck,' says Loren. 'To our knowledge, no other genetic disease has a "backup" gene with the potential to encode the missing protein.' No one knew how the SMN2 protein spigot worked exactly, or where one might find a wrench to open it up, but the dream of super-charging a gene that had been more or less useless throughout human history was compelling.

Loren, Dinakar, their family and friends created the SMA Foundation, using their own money to lobby for and directly fund research into SMA. Public awareness of SMA was almost non-existent, so they gave interviews to media outlets like Forbes, ABC Nightline and NBC. The Foundation recruited some of the country's leading neuroscientists and clinicians. They did everything they could to build political support for SMA research, coordinate disparate scientific activity, expand the available research tools, and persuade biotech and pharmaceutical companies to target the disease.

Yet just as the SMA Foundation took shape, so Arya went into physical decline. She remembers being able to walk, although she finally lost control of her limbs while still in kindergarten. She was given a walking frame, but soon had to graduate to a wheelchair. 'I gave my wheelchairs names,' she says with a smile. 'There was Daisy . . . and Miley, because I loved Hannah Montana. But my favourite was Vanessa, after Vanessa Hudgens in *High School Musical*.' It's impossible not to picture the little girl in the accessorised wheelchair, mesmerised by a dancing Hollywood star she could never hope to emulate.

'I would cry every night,' says Arya. 'I would ask my mother, "Why me? Did I do something wrong?" It felt so unfair.'

Loren decided not to spell out the inescapable progress of the disease for her daughter, so Arya had to work it out for herself. At first, she was mostly embarrassed by her weakening arms and legs, and she decided not to tell any of her young friends. Of course, it quickly became clear that concealing her condition was not a viable option. So her next strategy was to refuse to talk about it.

That wasn't helped by the publicity the Foundation was starting to generate, much of which inevitably involved Arya. 'I didn't want to be the face of SMA,' she says. 'I read those articles about me later and cried.'

For a long time, Arya wasn't allowed to Google herself. She did anyway, and then felt she couldn't talk to her parents about what she'd found out. 'Eventually, friends who'd also Googled me started coming to me and saying, "Oh my God, your life is so sad," or, "Oh my God, you're going to die." I was really mad with my parents about those articles.'

After a time, she came to accept that she would always be different from other people, although it was still painful to watch friends run around in the park or go on sleepovers from which she was excluded by a staircase or a broken elevator.

To understand how we're going to hack the *SMN2* gene and make it do something it has never done before, we need to peer inside the inner workings of a cell's genetic machinery. Proteins are essential building blocks in all our cells, and they are manufactured from blueprints encoded in our genes. We need to know how that happens to understand how the SMA nut was cracked.

Our genes are located on parallel strands of DNA, arranged in the iconic double helix first described by James Watson and Francis Crick in 1953. On any one strand of DNA, some parts encode useful proteins. These parts are known as *exons*. In between are lengths of seemingly useless genetic code called *introns*.*

Proteins cannot be directly manufactured from DNA. A 'messenger' is needed to carry the genetic information from the nucleus, where DNA is held, to the cell cytoplasm, where the amino acids that will be assembled into proteins are waiting. This messenger is called

* They are not in fact useless, as we shall see in Chapter 7.

RNA.* When a cell needs to build proteins, the parallel strands of DNA are temporarily separated, and the genetic code on one strand is *transcribed* to forge a complementary strand of RNA. It is this RNA which is used as the blueprint to manufacture a specific protein.

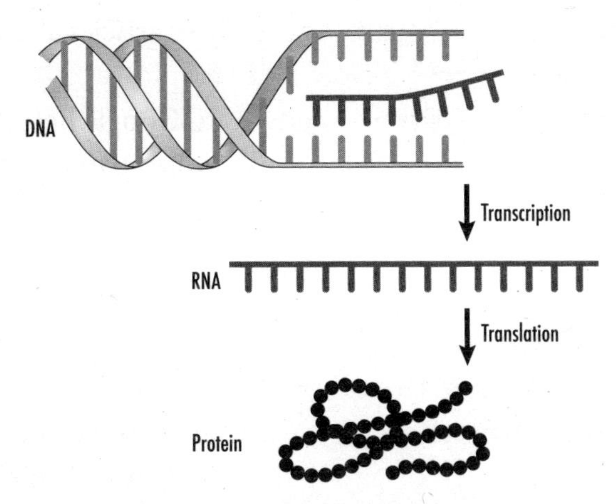

Figure 2. Genetic information flows from DNA to RNA to protein.

It's as if a factory owner keeps the master blueprints (the DNA) for their products locked in their office (the nucleus); any time the production line starts up, a supervisor has to go to the office and copy out the blueprints, then take that copy (the RNA) to the factory floor where the products (proteins) are made.

However, before protein production can start, all those apparently useless bits of genetic code – the introns – need to be stripped out of the RNA. This is done by *splicing.*

Picture a grizzled sailor with a piece of old line that's become frayed in the middle. Rather than discard the whole line, our sailor decides to cut out that frayed section and join the two remaining lengths together. He does this by *splicing*, unravelling the ends of each length and threading the strands together.

The *primary* strand of RNA consists of all the introns and exons transcribed from DNA. The splicing process removes the introns,

* More precisely, 'messenger RNA' (mRNA).

leaving just the exons joined together in a shorter strand of *mature* RNA. An assembly of proteins and 'small nuclear RNAs' called the *spliceosome* comes together to mediate the removal of each intron.

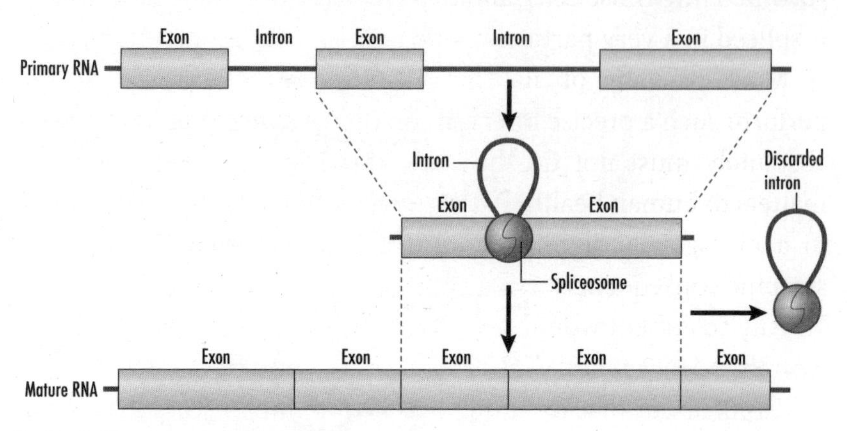

Figure 3. The process of RNA splicing.

You may ask why nature has come up with this byzantine process. Why on earth give us introns in the first place if it's just going to cut them out? The answer seems to lie in part in the greater diversity of proteins that can be generated from a relatively small number of genes: the splicing process can be varied to join different exons together, essentially making multiple products out of the same blueprint.

So proteins are manufactured only after *primary RNA* has been spliced to form the shorter *mature RNA* that is their final blueprint. Splicing is where *SMN1* and *SMN2* differ.

When the primary *SMN1* RNA is spliced to create mature RNA, a critical segment called Exon 7 is included. By contrast, when the primary *SMN2* RNA is spliced to create mature RNA, Exon 7 is usually left out. Without it, the resulting protein is not very functional. So, the challenge for anyone wanting to treat SMA was clear: fix the splicing of *SMN2* RNA to include Exon 7.

Remember, all of this is happening at such minuscule scale in the cell that no optical microscope could detect it. And SMN protein has to be produced in countless individual cells. We can't go in with a scalpel and splice that troublesome RNA by hand. Instead, we have to discover some substance that can be injected or ingested, that will find its way to every relevant cell, that will be absorbed into those cells, and that will change the way *SMN2* RNA is spliced in a very particular way.

Most crucially of all, this hypothetical substance that will perform such a precise intervention on the splicing of *SMN2* RNA absolutely must not do the same thing to any other gene that matters to human health. If another gene had its splicing disrupted in a biologically significant way, the patient could suffer catastrophic consequences.

Any treatment would need to be as *specific* as possible to Exon 7 on the *SMN2* gene.

Think about that for a moment. Whose heart wouldn't sink at the prospect of searching for a substance that could reach into our motor neurons and tweak a single genetic process at a single site within a single gene while leaving every other important gene untouched? Such a feat had never before been accomplished. Would you take on that challenge?

RNA splicing was discovered in 1977, but it remained only partly understood when Adrian Krainer took up a fellowship at New York's Cold Spring Harbor Laboratory (CSHL) in 1986. Born in Uruguay, Krainer had come to the United States on a college scholarship to study biochemistry, and had dedicated his PhD at Harvard to splicing research. He wanted to discover how the cutting and joining of RNA actually occurs.

A hub of splicing research, CSHL was the perfect place for Krainer to continue his investigations. By separating the contents of cell nuclei into fractions and then testing various combinations of those fractions, he was able to identify and purify some of the RNA-binding proteins that control splicing. These proteins would either promote exon *inclusion* or exon *skipping*; that is, they would determine what was kept in or left out of the final genetic code.

The most important of Krainer's early discoveries was a protein called SRSF1,* which must be present if splicing is to occur. Due to its particular shape, SRSF1 can bind only to certain configurations of RNA, so splicing will only occur where the right coding sequences are found. Over the next decade, Krainer and his colleagues uncovered many more of the secrets of our cells' splicing machinery.

In 1999, Krainer was invited to an SMA workshop at the National Institutes of Health (NIH). He was by then recognised as an expert on RNA splicing, and the organisers figured he could be encouraged to apply his expertise to the challenge of upgrading *SMN2*.

Krainer listened attentively as papers were presented on the omission of Exon 7 in *SMN2* splicing. He knew that mutations in or around an exon could cause exon skipping when SRSF1 could no longer bind to the RNA; he could see that the slight coding difference between the two *SMN* genes might in this way cause Exon 7 to be left out during the splicing process. But he would have to confirm it with experimental evidence. 'That was the moment of realisation,' he says. 'I thought, there is no way I'm *not* going to work on this disease.'

With a grant from the non-profit organisation Fight SMA, Krainer and postdoctoral fellow Luca Cartegni spent the next couple of years searching for the sequence on Exon 7 that SRSF1 should recognise. SRSF1 is an *activator* that should encourage inclusion of Exon 7 in splicing – but only if it can bind to the right sequence.

Meanwhile, a competing lab was working on the same problem, and was convinced that Krainer had it the wrong way around: they believed a *repressor* protein bound to Exon 7 was responsible for its skipping. It was a distraction that shook his confidence, Krainer admits, but the competition motivated him to accelerate his work.

Eventually, he identified the critical sequence on Exon 7 that should enable SRSF1 binding, and he showed that it was mutated

* Serine and arginine rich splicing factor 1.

in *SMN2*. In their genetic code, there is only a single letter difference between the *SMN1* and *SMN2* genes. At one particular position, *SMN2* has a *T* where *SMN1* has a *C*.* Such a tiny difference should hardly matter, you might think. But that one erroneous letter in *SMN2* is enough to disrupt the splicing process.

Now, the challenge was to find a substitute for SRSF1 that would work on the mutant sequence. Cartegni and Krainer designed a synthetic molecule consisting of an oligonucleotide (a short sequence of single-stranded nucleic acid†) attached to a peptide chain (mimicking part of SRSF1). They showed that it could bind to *SMN2*'s Exon 7 in a test tube, doing the splice-activating work that SRSF1 was failing to do. The higher the dose of this SRSF1 substitute, the more Exon 7 was included in the resulting mature *SMN2* RNA.

Krainer and Cartegni had proved the splicing of *SMN2* RNA could be fixed *in vitro*.[2] This exciting result suggested it should be possible to increase the production of SMN protein in SMA patients.

They published their results and patented the oligonucleotide-peptide molecule they'd designed. Their work caught the attention of C. Frank Bennett at Ionis Pharmaceuticals, who licensed the technology and collaborated with Krainer to develop a drug for SMA.

Together, Krainer and Bennett, with postdoctoral fellow Yimin Hua, found they could strip down the chimeric molecule to a shorter sequence of single-stranded nucleic acid that successfully corrected exon skipping. The new approach was an *antisense oligonucleotide* (ASO), a synthetic piece of chemically modified nucleic acid that held the mirror code to a sequence on Exon 7 of *SMN2* RNA. Later they showed that other ASOs, which mirrored code on the introns on either side of Exon 7, also worked to correct exon skipping. These acted by blocking sequences that would normally

* The genetic code is written in an 'alphabet' of four chemical *bases*: adenine (A), guanine (G), cytosine (C) and thymine (T).

† Nucleic acids are the group of chemical compounds that include DNA and RNA.

bind a protein that repressed splicing. It was one of these ASOs that proved the most effective drug.

In 2008, Krainer and Bennett tested their lead ASO in mice that had been genetically modified to mimic a patient with SMA (with a human *SMN2* gene). They showed that these transgenic mice synthesised more human SMN protein when treated with the ASO. They had proved that SMA could be beaten, at least in animal models.[3]

Adrian Krainer first met Loren Eng at an SMA event in the early 2000s. He was impressed by the way she bridged the gap between different SMA organisations. 'She brought peace,' he says. Her non-profit also provided Krainer and Bennett with funding to help identify their ASO.

But the SMA Foundation was not content to bet on only one research lead. Krainer and Bennett's oligonucleotide approach had some advantages: it was incredibly specific, as the oligonucleotide would only bind to the unique RNA code next to *SMN2*'s Exon 7. But it also had drawbacks: an ASO would not be able to cross the blood-brain barrier, which separates the central nervous system from the rest of the body, so it would have to be injected directly into the fluid surrounding the spine, where the nuclei of motor neurons are located. The SMA Foundation wanted to encourage parallel research into simple small molecule chemicals that could be swallowed in a pill or capsule. For that, they turned to a young New Jersey biotech company called PTC Therapeutics.

Stuart Peltz had been fascinated by RNA since grad school, in part because it had been so long overshadowed by its more famous cousin, DNA. Yet Peltz could see that, as the intermediary in the protein-production process, RNA might be exploited to control what genes actually do. This was a novel concept at the time. In earlier decades, scientists had assumed that whatever was encoded in DNA was exactly what you got; more recently, it was becoming clear that genes could be turned off and on, and the way proteins were synthesised could be influenced by environmental factors.

For Peltz, harnessing the power of RNA control processes might unlock all sorts of therapeutic possibilities.

Peltz established PTC Therapeutics in 1998 to take on rare genetic diseases through RNA modification. PTC stands for 'Post-Transcriptional Control', reflecting his company's focus on controlling gene expression at the RNA level.

By 2003, PTC had successfully developed its first drug, a small molecule targeting a genetic mutation implicated in Duchenne muscular dystrophy (DMD), another fatal neuromuscular disorder. At the time, there were no effective treatments targeting the underlying cause of the disease. Building on this work, Peltz and his team were already looking into spinal muscular atrophy when the SMA Foundation came calling.

Fixing *SMN2* by correcting its RNA splicing would be a challenging proposition, Peltz acknowledged. No one knew what kind of chemical might modulate *SMN2* RNA splicing, and there were no precedents to suggest the idea could even work. The PTC team would need to break new ground. But they began the way pharmaceutical companies have always begun: they used trial and error on a grand scale to find a chemical that showed some kind of positive effect.

Since its inception six years earlier, PTC had built up a diverse library of around 300,000 chemical compounds. To discover whether any of them could increase production of functional SMN protein, the team created a high-throughput *in vitro* assay. This is a centrepiece of modern drug discovery, a mechanical system that can test hundreds of thousands of different chemicals against a biological disease target. It requires considerable engineering precision to deliver exactly the right dose of each chemical to a small well containing the target, along with some creative thinking on how to detect a positive response in any one well.

'We couldn't always afford the newest gadgets in the lab,' says Peltz, 'but we were scrappy back in those days, and the team knew how to think outside the box.' They bought an old Coke machine on eBay and repurposed it with their own robotics to store and load chilled sample plates for the assay.

It was a marathon process for the fifty chemists on the project. When a particular compound was found to elicit a slight increase in functional SMN protein, the team would test every related chemical in their library to see if any of them could do better. Then they would turn to the open market, buying any related chemicals they didn't already possess. Finally, when they had tried every chemical commercially available, the chemists synthesised brand new compounds. Each chemist was able to create between one and three completely original chemicals per week.

'You can think of each molecule having a core with some jewellery around it,' explains Peltz. 'We would keep changing the jewellery until the potency of the molecule got a bit better.'

Such molecular engineering would lead to incremental increases in the amount of functional SMN protein produced in the petri dish assays. But eventually tweaking the 'jewellery' would yield no further improvements, and the team would have to rethink the core of the molecule. It was this iterative process, balancing small improvements against setbacks and dead ends, that eventually led to a breakthrough. Gradually, step by arduous step, PTC's chemists modified their most promising molecules until they had a candidate that really boosted SMN protein production.

But this hard-won efficacy was only half the battle. Next, they had to work on all the other drug qualities of the molecule, such as its bioavailability (how easily it is absorbed from gut into bloodstream), its pharmacokinetics (what happens to it in the blood), and its possible side effects.

It took PTC seven years to inch their way towards a set of molecules that seemed both effective and safe. 'I don't believe in failing fast,' Peltz says with a smile.

Alongside the *in vitro* assays in petri dishes, PTC tested its most promising molecules *in vivo*, in mice that had been genetically modified to develop SMA. Normally, such transgenic mice would wither and die within five days of birth. PTC showed their lead candidate molecule could give these mice something close to a normal lifespan. When mature *SMN2* RNA from the treated mice was sequenced, Exon 7 was found to be present in most strands: it was no longer being cut out during the splicing process. This was a watershed achievement, a major breakthrough in the field of small-molecule splicing modulation.[4]

PTC had proved the basic concept at the heart of the *SMN2* dream: a simple chemical could modify *SMN2* RNA splicing, and so increase functional SMN protein production enough to let an SMA mouse develop with little, if any, impairment.

A start-up like PTC Therapeutics, even with lavish venture capital funding, can only afford to work on one or two programmes of this size, and it would take many years to see any return on their investment. Despite the promise of their DMD drug, the company still wasn't generating any income other than research grants and some collaboration income. By the time their SMA mice were living long enough to reproduce, PTC had burnt through hundreds of millions of dollars. 'We hit a wall in 2011,' admits Peltz. 'I couldn't convince anyone to give us any more money.'

Pharmaceuticals are not like other technological innovations. You can't just invent something and start selling it the next day. Where Silicon Valley innovators can knock up a new computer or drone in their garage and take a few prototypes to sell at a trade show, pharmaceutical companies have to go through years of clinical trials and regulatory submissions before a medicine can be sold to patients. Such trials come with a very hefty price tag, so it is rare for smaller companies to take their innovations all the way to market on their own. There comes a point where they need to hand over to a bigger player, a partner with the experience, expertise and resources to take their compound through clinical trials, regulatory approval and commercial launch.

For Krainer and Bennett, that partner was Biogen. For PTC, it was the Swiss giant, Roche.

Luca Santarelli was driving across the Alps when he got the call about PTC's breakthrough. As head of neuroscience at Roche's Pharma Research and Early Development (pRED) unit, he lived and worked in Basel, Switzerland, but his family home was on the other side of the mountains, in Turin, Italy. Santarelli made the four-hour-plus drive most weekends in his Audi A5 coupé. The call was from his heads of neuroscience research and early development, respectively. Both were clearly excited.

Paulo Fontoura, in charge of early development, said, 'We need to get this! It has the potential to become an important medicine.' He'd been impressed by the elegance of the approach PTC had taken. At this point, in 2011, the concept of modifying genetic splicing in humans was still science fiction. No one had ever made a small molecule drug that acted on RNA like that. Yet the results from the animal models were clear: PTC had three candidate compounds that seemed amply potent.

Fontoura had once met a patient with SMA from South Africa, a young man in a wheelchair, and he was struck by his ambition: the man wanted to be an astronaut. 'As long as you can stop this disease from killing me,' the young man had told him, 'I can do the rest.' Now Roche had the opportunity to play a part in that patient's dream.

Santarelli, a psychiatrist and neurobiologist by training, was astonished by PTC's achievement. 'It was completely unheard of, at that time. I was immediately sold. I said, "Let's go get it."'

Later, he watched the video of SMA mice given new life by one of PTC's molecules. 'These mice that were dragging their feet and dying, they made them survive. It was one of the most impressive things I'd ever seen.'

Other pharma companies were also interested in PTC's breakthrough, so Roche would have to move fast. And there was a twist: as part of the funding deal the SMA Foundation had made with PTC Therapeutics, Loren Eng would have the right to attend every joint steering committee if the resulting molecule was sub-licensed to a pharma company.

'We would be developing a drug with the mother of a child with SMA sitting at the table,' says Santarelli. 'This was the most unprecedented situation, but also very exciting.'

Santarelli was headed to New York a few days later, so he arranged a meeting with Loren in the lobby of the Soho Grand Hotel. 'I was immediately impressed,' he says. 'She had this incredible sense of purpose.'

Comfortable that the technology was sound and the three organisations were all on the same page, Santarelli and his boss made the deal. In November 2011, ten years after Arya Singh was diagnosed, Roche bought an exclusive worldwide licence to PTC's SMA programme. PTC provided Roche with three lead compounds and a collection of around four thousand other chemicals they had identified as having *SMN2* RNA-splicing potential.

'It was the highlight of my professional career,' says Santarelli. 'We scientists are driven by intellectual curiosity and motivated by the idea of helping patients. But the patients are usually a remote concept. With Loren so closely involved, this was much more emotional.'

PTC had identified a number of molecules that worked in a mouse model of the disease, but could any of them be a real drug? There is a big difference between a molecule that shows proof of concept in the laboratory and a medicine that can be prescribed to patients. One key question was safety. A molecule that modified RNA splicing at one genetic site was likely to have an effect on other sites too. That might not matter if the affected genes determined hair colour or nose shape, but anything more physiologically significant could have toxic consequences.

Roche toxicology project leader Lutz Müller was a member of the early drug development team, and he uncovered several challenges with the three lead compounds from PTC. The most advanced compound was phototoxic, meaning it could potentially render patients susceptible to sunburn. It sometimes caused DNA mutations. It could potentially affect the heart and other organs. And it was unstable at room temperature, meaning it might need special formulation and storage. A review of the other two compounds revealed similarly critical concerns.

These molecules might be efficacious, but would they have a real chance of becoming a medicine?

Another key member of the team was Hasane Ratni, a medicinal chemist. It was his job to assess and refine the chemical compounds under investigation. His profession combines deep scientific knowledge with an artist's instinct.

A molecule is a set of atoms bonded together in a particular way. The water molecule comprises two hydrogen atoms bonded to an oxygen atom. The ethanol molecule is made up of two carbon atoms and one oxygen atom bonded to six hydrogen atoms. Individual atoms or *groups* of atoms can be added to, or removed from, a molecule to change its attributes.

It's basically Lego.

Ratni's job was to arrange the Lego blocks to create the best molecule.

'We have a kind of toolbox,' he explains of medicinal chemists. 'Techniques we've developed over the years to break off an oxygen atom here or attach a methyl group there.' There's a dash of trial and error, but in general, says Ratni, once you know what atom you want where, it's fairly straightforward to put it in place.

The difficult bit is deciding how the Lego blocks should be arranged.

Some parts of the lead molecule were clearly interacting with the splice sites around Exon 7 on *SMN2* RNA, and these had to be preserved. It was the other parts – the parts that made the molecule unstable and toxic – that had to go. Those were the Lego blocks – or *fragments*, as he would call them – that Ratni needed to switch out.

But which parts of the molecule were the problematic fragments, and what could he safely put in their place?

Hasane Ratni has dedicated his entire career to drug discovery. Still youthful, he might be expected to have all kinds of outside interests, but while he admits to the occasional football game and restaurant meal, 'You never really disconnect.' Making drugs is everything for him. He remembers the intense joy and motivation he felt when he heard that Michael Sofia had invented a cure for hepatitis C. 'I was super-impressed. It was like, *Wow, oh my God!* A serious disease can just disappear if a chemist builds the right molecule.'

With a team of chemists and toxicologists, Ratni got to work on the lead molecule. Fixing the instability problem was easy. 'That was just rational design,' he says. 'We could see which fragment of the molecule was causing it to break down, so we replaced it.'

The phototoxicity problem proved tougher. Ratni reduced the unwanted effect by switching out another fragment. The molecule remained somewhat phototoxic at high dosage levels, but at the dosage expected to be administered in humans the sunburn risk would be minimal.

The biggest issue, however, was the molecule's tendency to make DNA mutate, leading to a risk of cancer. What could be causing that? Ratni suspected the double aromatic ring. He cut it out of the molecular design and the team synthesised the new compound. It wasn't mutagenic any more . . . but it didn't modify RNA splicing either. He had fixed the toxicity problem but destroyed the molecule's effectiveness.

He tried another fix. The same thing happened. And again. And again.

So it went for nine long months. Every time the chemistry team tried another tweak to make the molecule non-mutagenic, they rendered it ineffective.

The delay was draining. Ratni remembers one occasion, a visit to London in 2012, when the Roche team met the parents of babies with the most severe form of the disease. The babies he saw in photographs and videos would never be able to sit up. They would develop breathing problems and would become partly paralysed. One couple came up to him and thanked him profusely for the work he was doing for their daughter. He couldn't answer them; he knew their little girl would be dead in less than two years, and there was no chance Roche could deliver a viable medicine in time to save her life.

Ratni and his colleagues were closing in on a conceptual structure for a safe and efficacious molecule, and now they turned back to the compounds of interest that PTC had supplied as part of the licence agreement. None had so far been selected for clinical development, but might there be a hidden jewel among those four thousand PTC compounds? The chemistry team went through them one by one, searching for a molecule that resembled the conceptual structure they had in mind. It was a slow, dogged process, but finally they found one that threaded the needle: it offered the splicing potency PTC had achieved with its lead compounds, and it passed all required toxicity tests.

The new lead candidate was given an official label: RG7800. It was stable. It showed no problematic toxicity at efficacious doses in initial animal trials. It was ready to test in humans.

The first step was a study of RG7800 in healthy volunteers. The development team measured blood concentrations of the drug after dosing, as well as the impact of various dose levels on the inclusion of Exon 7 in *SMN2* RNA. They found that RG7800 could double levels of full-length, functional *SMN2* with Exon 7, without causing significant side effects. RG7800 had performed exactly as intended. It was deemed a categorical success.[5]

Everyone at Roche, at PTC, and at the SMA Foundation was elated. It was that rare moment that drug developers long for – proof of mechanism for a completely new drug concept. For Ratni and his colleagues, it felt like the pinnacle of their careers.

Roche's SMA programme had been codenamed 'Project Nemo' after the little cartoon fish with the deficient fin. Now, with a nod to the school of fish that literally point the way in the Pixar movie, the clinical trial of RG7800 in SMA patients was named MOONFISH.

To prove the molecule's efficacy, the Roche development team would ultimately need to show that babies with the most severe form of SMA lived longer and could do more on a regimen of RG7800. But clinical trials involving babies are, as you can imagine, a complex undertaking, so the first MOONFISH study was conducted in older patients with a less severe form of the disease. The investigators needed to see increased production of full-length mature *SMN2* RNA and higher concentrations of functional SMN protein in dosed patients. They would also look for side effects and monitor how the molecule behaved in the human body.

There was a long way to go, everyone knew, but that proof of mechanism in the initial study motivated them all. Success seemed almost assured.

Lutz Müller remembers vividly the day pathologist Annika Herrmann came into his office and announced, 'We have a problem.'

At first glance, Müller is the picture of the stereotypical serious German scientist. He began his long career in West Berlin, as it was called then, working for the Federal Ministry of Health, where he reviewed pharmaceutical product safety data and helped write global guidelines for drug development. He was tempted out of government service only when his office was relocated to Bonn, a city he did not love. Instead, he moved to Basel to take a job with Novartis, and from there skipped across town to join Roche.

But behind the serious front is a wilder side. Müller was one of those young Berliners who took a hammer to the Wall in 1989. Now a grandfather, his house is decorated in bold, bright colours, and from his home office window he looks out on a Jaguar E-type roadster in primrose yellow, the pride of his vintage car collection. He even has special yellow shoes for driving it at speed around the Black Forest. The non-conformity extends to his work: 'I'm seen as the cowboy of the project leaders,' he says. 'I tend to ignore good advice.' In Müller's view, if you know what you're doing it's OK to challenge boundaries, even bend the rules sometimes. Those who aren't willing to go off-piste once in a while aren't likely to make the big scientific breakthroughs.

But now his pathologist had brought him a problem he couldn't simply accelerate past.

Herrmann had been reviewing tissue samples from a 39-week animal toxicology study of RG7800. The two-week animal studies that preceded MOONFISH had thrown up no red flags. This much longer 'chronic toxicology study' felt almost a formality – just to check nothing turned up in monkeys exposed to longer courses of the drug.

Müller took one look at the histopathology image she'd laid in front of him and saw it immediately.

'These are photoreceptors?' he checked.

Herrmann nodded silently.

The tissue sample came from the retina of one of the monkeys dosed with RG7800. The photoreceptors – cells critical to vision – were damaged. Permanently.

Müller couldn't believe it. He had worked in toxicology for thirty-seven years and had never seen anything like it. The ophthalmology review was a standard part of the drug safety testing process, but no molecule he had previously worked on had ever shown such a result.

After all the excitement and celebration of the initial results, Müller was faced with one of the worst duties of his career. He had to recommend to the team that the study be stopped immediately due to safety concerns. None of the patients in MOONFISH had suffered retinal damage, and the animal finding might not be relevant to humans, but the theoretical risk was just too high. The enrolled patients were told that their potential treatment was being withdrawn. Later, their blood samples would provide confirmation that production of full-length mature *SMN2* RNA had increased and that functional SMN protein levels were significantly elevated.

Nevertheless, RG7800 was history.

When Hasane Ratni heard the news, he was devastated. 'It felt like you've won the World Cup and then someone snatches it away.' He asked himself if there was something that could be done to keep the trial going – some tweak he could make to the molecule to eliminate the retinal toxicity. But an in-depth review of the toxicology findings did not suggest any quick fix.

He would have to go back to the drawing board.

Equally devastated were Loren Eng and the team at the SMA Foundation. Arya's torments were mounting. She suffered orthopaedic deformities and hip dislocations as a result of her muscle weakness and had to wear leg braces to prevent further damage. She was unable to turn herself in bed, so had to have someone at her side twenty-four hours a day. She could no longer stand, crawl or even lift her head. She was losing the ability to raise her arms. She was in constant, severe pain. And she had to endure the thoughtless cruelties of classmates and the ignorant remarks of strangers. 'Over the years, Arya has suffered everything,' says Loren.

Every day, Arya had to undergo three hours of physical therapy, thirty minutes of respiratory exercises and twenty minutes of stretches. Each physical task took her two or three times longer than anyone else. 'SMA adds four hours of work to your day,' she says. Clothing had to be chosen carefully for any excursion as she wouldn't be able to take off layers: too much and she'd sweat heavily; too little and she'd risk being hospitalised with pneumonia. A simple chest infection is extremely dangerous for SMA patients as they struggle to cough unaided and so may be unable to clear excessive mucus from their lungs. Pulmonary failure is the leading cause of death in SMA children. Twice, Arya had suffered a collapsed lung.

'My darkest time was aged twelve.' She had to undergo terrible surgeries to rectify the consequences of her wasting body. An operation to fix her hip left her with one leg four inches shorter than the other.

'When I awoke from sedation, I attempted to write, "What is happening to me?" (The intubation made it impossible to talk.) When the nurse couldn't read my handwriting, I knew something was really wrong. I was known for my penmanship! I couldn't stop crying. Afterwards, I spent days feeling hopeless, just staring at the changing numbers and the pattern of the ECG machine.'

The termination of the MOONFISH trial, disappointing as it was, at least gave Ratni, Müller and the team some breathing space after what had been a high-pressured initial design phase. In parallel to the work on RG7800, the team had been developing a number of backup molecules, just in case the lead candidate failed. Now, they returned to those backup molecules and, through a painstaking selection process that lasted several months, they identified one that had a lower risk of retinal toxicity while still maintaining potency.

The new molecule, RG7916, was submitted for the chronic toxicology study.

To the intense relief of the whole team, it passed. Roche could finally resume testing in patients.*

Paulo Fontoura had by then become head of late-stage neuroscience clinical development, and he quickly assembled a new team to take RG7916 into clinical trials. Their objectives were ambitious: 'We wanted to bring this new potential treatment to all types of SMA patients, not just the most severe cases,' he says. 'We wanted to run the trials globally, so that if we were successful there would be minimal lag in patients everywhere having access to the treatment. And we needed to do it super-fast.'

The team, led by experienced drug developer Jean-Paul Pfefen, set up two global trials to treat patients from as young as two months up to twenty-five years old. They employed seamless adaptive trial designs, meaning they reviewed the data while the trials were still in progress and made changes in real time to speed up the overall process. The first safety trials started in January 2016, with two pivotal trials following later in the year. By the end of 2018, the data confirmed everything Roche, PTC and the SMA Foundation had hoped for: RG7916 was efficacious in all types of SMA patients. It was given a name: *risdiplam*. On 7

* Roche went on to monitor for retinal toxicity extensively in clinical trials, and no ophthalmological issues were found.

August 2020, the new drug was approved by the US Food and Drug Administration (FDA).

In the meantime, two other treatments for SMA were approved by the FDA. The first was *nusinersen*,* the antisense oligonucleotide from Ionis Pharmaceuticals and Biogen based on Adrian Krainer's work. The second was *onasemnogene abeparvovec*.† This treatment emerged from work led by neurobiologist Brian Kaspar at Nationwide Children's Hospital in Columbus, Ohio.

Kaspar tackled SMA quite differently to Krainer and Peltz, using an exciting new form of medicine: gene therapy.

We'll have more to say about gene therapy in Chapter 7, but the basic idea is to replace a dysfunctional gene in the patient's DNA with a functional one. In this case, Kaspar used a specially engineered virus as a vector to deliver genetic material (called a cDNA) encoding a replacement *SMN1* gene to the patient's nerve cells. *SMN1*, remember, is the gene missing in people with SMA.

Nationwide Children's Hospital licensed the technology to a biotech start-up, AveXis, in 2013, which took it into clinical trials.[6] AveXis was acquired by Novartis in 2018, and the following year onasemnogene abeparvovec was approved.

Risdiplam was thus the third treatment for SMA launched in just under four years. For the SMA Foundation, the success of these three treatments – two of which they had directly funded, and all of which they had supported with tools and scientific insight – was an extraordinary achievement.‡

Hasane Ratni and Lutz Müller marked the FDA approval of risdiplam, in the middle of the Covid-19 pandemic, in a characteristically understated way: on a video call shortly after the

* Approved on 23 December 2016 for SMA patients of all ages and types.

† Approved on 24 May 2019 for SMA patients under two years of age.

‡ AveXis benefited from a major investment from venture capital firm Deerfield under the guidance of partner Jonathan Leff, who sits on the board of the SMA Foundation.

announcement, they both appeared, grinning, in T-shirts emblazoned with the new medicine's brand name.

'Eight years of work for a T-shirt,' laughed Ratni.

'I am never taking mine off,' declared Müller, mock-serious. 'Not till it rots off me.'

Paulo Fontoura looks back on those days with feelings of elation and pride. 'This is all we could have hoped for, and it was only possible because we had such an amazing team and great partners,' he says. 'These are the projects that define your career.'

Risdiplam can now be prescribed to patients diagnosed via neonatal screening from the moment they are born.

For me, the success of three SMA programmes in such close succession is a remarkable example of human ingenuity. All three draw on the genetic understanding of the disease, but they address the problem in quite different ways, using different technological advances. Together, they demonstrate how new scientific knowledge can quickly unlock innovation.

For Loren Eng, understandably, the success of the three programmes is bittersweet. She is overjoyed that the SMA Foundation has helped deliver three viable treatments, but distraught that they came a bit too late for her daughter. 'We have stabilised Arya, but we have not found a treatment that will give back function for my child,' she acknowledges, after twenty years of searching. She and the SMA Foundation are now investing heavily in regenerative medicine for neural circuits and muscle. Meanwhile, though, Arya is still in a wheelchair.

Arya is more optimistic. She now accepts the physical limitations on her life with remarkable grace, and is determined to excel in all other areas. Her strong performance in high school earned her offers from Stanford and Barnard universities, but she chose to attend Yale. She graduated at the top of her class and was elected commencement speaker. She led numerous college clubs and was president of Yale's largest student organisation, the Legal Aid Society. She was also a finalist for the Rhodes Scholarship and won Yale's two major prizes – for academics and character.

'SMA has made me physically weaker, but I decided early on that it should make me mentally stronger,' she says. Following her mother's example, she wants to make a difference in the world.

And in what field does Arya Singh plan to build her career?

Health policy, she suggested ... or perhaps even drug development.

CHAPTER 2

Lung Cancer in Never Smokers

Condition: Lung Cancer
Target: EGFR
Innovation: Osimertinib [oh-si-MER-ti-nib]
Company: AstraZeneca

A SEPIA-TINTED PHOTOGRAPH OF THE LEVI-MONTALCINI family, taken around 1912, shows young Rita Levi-Montalcini in a starchy white dress, clutching a straw hat several times larger than her head.[1] She looks properly cross. From the beginning, she did not fit into the traditional mould Adamo Levi had conceived of for his daughter. 'My experience in childhood and adolescence of the subordinate role played by the female in a society run entirely by men had convinced me that I was not cut out to be a wife,' she wrote in her memoir. 'Babies did not attract me.'[2]

Having initially denied his daughters anything more than a basic education – a high school for girls that taught no biology – Levi eventually folded and allowed Rita to apply to medical school at the University of Turin. After an accelerated effort to catch up on all the study missed, she was accepted into the course and graduated with distinction in 1936. By then, Rita Levi-Montalcini had become fascinated with biological research, and she stayed in academia, studying neurology and psychology until 1938, when

Benito Mussolini's *Leggi Razziali* banned Jewish people from higher education.

Undeterred, Levi-Montalcini continued her studies in Brussels. Anticipating the German invasion of Belgium in 1940, she returned to Turin and set up a laboratory in her bedroom. The question that preoccupied her through the first years of World War II was how nerve cells emerging from an embryonic nervous system find the muscles or other elements of the body to which they need to connect.[3] Her research subjects were chicken embryos scooped out of eggs from the local farm. Using sewing needles for scalpels, an ophthalmologist's scissors, and tweezers from a watchmaker, she teased out the embryonic nerve cells under a microscope to learn how they grow.

In 1943, Italy surrendered to the Allies and was swiftly invaded by its former Axis partner, Germany. With Nazi forces now in control of large parts of the country, the situation for Italian Jews turned from bleak to disastrous. Thousands were rounded up and sent to death camps.

The Levi-Montalcini family fled from Turin to Florence, but failed to stay ahead of the invading forces. Instead, they assumed false identities and sought refuge with non-Jewish friends. Day after day they hid from the occupying Germans. Meanwhile, Allied troops were fighting their way up from the south of Italy, taking Rome in June 1944 and reaching Florence two months later. At last, the city was free and the Levi-Montalcinis could breathe again.

After the war, Rita Levi-Montalcini received a surprising invitation from the United States. The eminent embryologist Viktor Hamburger had read a paper she had co-written on embryonic nerve cell death. Impressed by her research and insight, he offered her a place in his lab at Washington University in St Louis, Missouri.

Levi-Montalcini described those years as, 'one of the most intense periods of my life in which moments of enthusiasm and despair alternated with the regularity of a biological cycle'.[4]

Working some of the time with Hamburger in St Louis and some of the time at a friend's laboratory in Rio de Janeiro, Levi-Montalcini discovered in the 1950s that a soluble substance produced by mouse tumours promoted rapid growth of certain nerve cells in chicken embryos.

She partnered with a young Washington University colleague, biochemist Stanley Cohen, to investigate this mysterious substance further. They identified a previously unknown protein they called *nerve growth factor* (a *factor* is a substance involved in a biochemical reaction). It would be the first of many proteins found to stimulate cell growth.

An amateur clarinet player and white-water canoeist from New York, Cohen was the model of the absent-minded scientist too absorbed in his work to notice worldly concerns such as the lighted pipe he'd stuffed in a pocket or the April Fool's jokes played by his colleagues. For his PhD, he had burrowed deep into the metabolism of earthworms at the University of Michigan: 'I remember spending my nights collecting over 5,000 earthworms from the University campus green,' he recalled.[5] Cohen later said of his collaboration with Levi-Montalcini, 'On our own we were good and competent. Together we were marvellous.'[6]

Cohen moved on to a faculty position at Vanderbilt University in Nashville, Tennessee, where he discovered a second growth factor in 1962. He made the discovery by extracting a substance from the salivary glands of newborn mice and showing that injecting it into other young mice accelerated the first opening of eyelids and the appearance of teeth, due to the stimulation of epidermal cell growth and keratinisation.[7] Named *epidermal growth factor*, this protein was subsequently found to play a vital role in skin and gut tissue development and repair.

The discovery of growth factors begged the logical question: how does a growth factor – a protein circulating in blood – cause cells to divide and multiply? Researchers like Ira Pastan and Jesse Roth speculated there must be some transmission mechanism to convey that growth *message* (as opposed to the protein itself) through the cell membrane and into the cell. That is, there must be some kind of *receptor*, a molecule in the cell membrane that is activated by the growth factor.

It was a small team in Cohen's lab at Vanderbilt that discovered the *epidermal growth factor receptor*, or EGFR, in human cells in 1975.[8] EGFR is a protein that stretches from one side of the cell membrane to the other. When epidermal growth factor binds to EGFR on the *outside* of the cell, it triggers a series of signals *inside* the cell that lead to cell division.

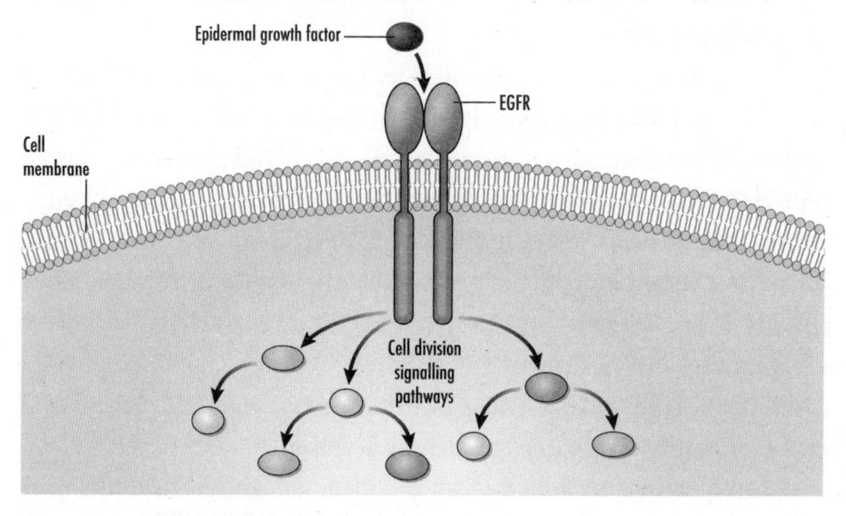

Figure 4. The EGFR signalling cascade.

Several years later, researchers found that some cancer cells *overexpress* EGFR, meaning they have an excessive number of these receptors. While normal cells may have a few thousand EGFRs in the membrane, some cancer cells have hundreds of thousands. This makes them exquisitely sensitive to epidermal growth factor, and in some cases these receptors are activated even in the absence of growth factor, leading to uncontrolled cell division. This pivotal discovery awakened the idea that cancer – which is characterised by excessive cell proliferation – could arise due to abnormalities in cell signalling involving EGFR.

So it was that EGFR became a primary target for drug hunters in search of new cancer therapies in the 1980s. If there were too many EGFRs on cancer cells, they reasoned, then a drug that blocked EGFR activity might slow or stop cell proliferation. Normal cells with fewer EGFRs would be less affected, it was hoped.

Even by the time I started seeing patients in the late 1990s/early 2000s, our primary tool for treating most cancers was still chemotherapy, and the side effects of those drugs could be daunting. Cisplatin could cause such severe nausea and vomiting that oncologists had to co-administer anti-nausea medicines. The taxanes, like paclitaxel and docetaxel, could cause neuropathy – numbness and tingling in the hands and feet. Every visit in the clinic, patients on these drugs were asked if they had trouble buttoning their shirts. If it got to that point, the taxane would be discontinued. All these chemo drugs took out white blood cells, red blood cells and platelets, leading potentially to increased susceptibility to infection, fatigue and bleeding problems, respectively. An EGFR-targeting medicine that left normal cells untouched would be a much more precise – and therefore more tolerable – cancer therapeutic.

Cohen and Levi-Montalcini were awarded the Nobel Prize in 1986 for the discovery of the two growth factors. The Nobel Assembly expected their discovery to 'increase our understanding of many disease states such as developmental malformations, degenerative changes in senile dementia, delayed wound healing and tumour diseases.'

Yet neither scientist had set out to address cancer ('tumour diseases'). As Graham Carpenter, a biochemist on the Vanderbilt team, said of Stanley Cohen, 'He understood [epidermal growth factor's] biological importance. But we did not have any idea that this would extend to cancer biology in a major way.'[9] The life-saving innovation that is the subject of this chapter was an unintended by-product of their fundamental research.

'Many new things are found by accident,' Cohen said later. 'If you're prepared to see the accident, you can find it.'[10]

* * *

The first two oral small molecule drugs designed to block EGFR signalling entered clinical trials in the late 1990s and early 2000s: gefitinib from AstraZeneca and erlotinib from OSI Pharmaceuticals.* At the time, I was a medical oncology fellow in training at Memorial Sloan Kettering Cancer Center (MSKCC) in New York City. My clinical mentors were expert oncologists Mark Kris and Vince Miller, and my focus was lung cancer.

Lung cancer is the number one cause of cancer death in the United States.† Back in 2000, patients with metastatic disease (where the cancer has spread outside the lung) usually lived for less than a year. Without treatment, median life expectancy from initial diagnosis was six months. With treatment, the patient might live an extra two to four months. The only treatment we could offer was 'modern chemotherapy', and we had no way to predict which patients it would benefit.

When starting first-line treatment,‡ I would explain to patients: 'I'm going to give you two cycles of chemotherapy, which involves two different medicines. We have a one-third chance your cancer will shrink. There's a one-third chance it won't change. And there's a one-third chance your cancer will keep growing. We'll only find out which it's going to be after a new CT scan. And you're likely to suffer side effects that include a risk of hospitalisation, debilitation and even death.' It floored me to have to present patients with such poor odds. Why couldn't we offer them anything better?

AstraZeneca and OSI had high hopes for their EGFR-targeting medicines. EGFR was reported to be overexpressed in many different cancer types, so they anticipated their molecules could become the next blockbuster drugs: targeted therapies effective against a wide range of cancers with minimal side effects. And at first, it seemed their hopes were justified.

* Cetuximab, an anti-EGFR antibody, was developed a little earlier, and is used to treat certain types of colorectal and head and neck cancers.

† This chapter concerns non-small cell lung cancer, the most common type.

‡ 'First-line treatment' is the initial medication selected for a patient; further lines of treatment are tried if/when the first line proves/becomes ineffective.

One of the patients I saw with Dr Kris was a middle-aged woman from upstate New York. Lung cancer is intimately associated with a history of smoking cigarettes, but she had never smoked. Because her disease was widespread and resistant to chemotherapy, her doctor had told her he had nothing left to offer; he recommended going to a hospice. Instead, she came to us. She arrived in a wheelchair, on supplemental oxygen, in a terrible state. X-rays of her lungs showed big areas of what I'll call 'white fluff' – the cancer.

We prescribed gefitinib.

Three days later, she called us and declared, 'I feel like a new woman!' She had stopped using oxygen. She was out of her wheelchair. She was active again. And she had minimal side effects from her treatment. We got new X-rays two days later. Remarkably, the white fluff was all gone: her lungs were clear. I had never seen anything like it. Her cancer stayed at bay for more than a year – longer than most patients with metastatic lung cancer lived at that time.

In medicine, we call such miracles the Lazarus effect.* If there was any one moment that ignited my passion for drug discovery, it was surely that. Other oncologists witnessed this miracle in some of their patients, too.

Yet in most patients with lung cancer, gefitinib did nothing at all.

It was a devastating disappointment – for us, for many of our patients, and for AstraZeneca. Across most of the different cancers included in the clinical trial, gefitinib showed no benefit. Only in a small proportion of patients with lung cancer (12–18 per cent) was there significant tumour shrinkage. Almost none of them were smokers, an unusual thing among lung cancer patients. We saw a similar pattern with the other EGFR inhibitor, erlotinib.

What was it about this small subset of patients that made these EGFR inhibitors so effective in them?

* * *

* After the Bible story in which Jesus resurrects Lazarus of Bethany.

Before we try to answer that riddle, let's take a moment to understand how these medicines work.

EGFR is a complex molecule made up of several component parts, including a *tyrosine kinase*. Tyrosine kinases are involved in many cell functions, including cell signalling, growth and division. More generally, *kinases* are enzymes that *phosphorylate* other molecules. When a kinase adds a phosphate group to a molecule, it changes that molecule's behaviour in some way. In the case of EGFR, it actually phosphorylates *itself* when triggered by epidermal growth factor. Phosphorylated EGFR then starts a chain reaction of signals inside the cell that leads to cell division and proliferation.

Kinases obtain the phosphate group they need for phosphorylation from ATP, an abundant molecule that is used in our bodies primarily as an energy store.* For phosphorylation to occur, an ATP molecule must attach to a *binding site* on the kinase. The most tried-and-tested way to inhibit a kinase is to block that binding site with a drug. That is what gefitinib and erlotinib do. By blocking ATP binding, they inhibit EGFR and so shut down signalling.

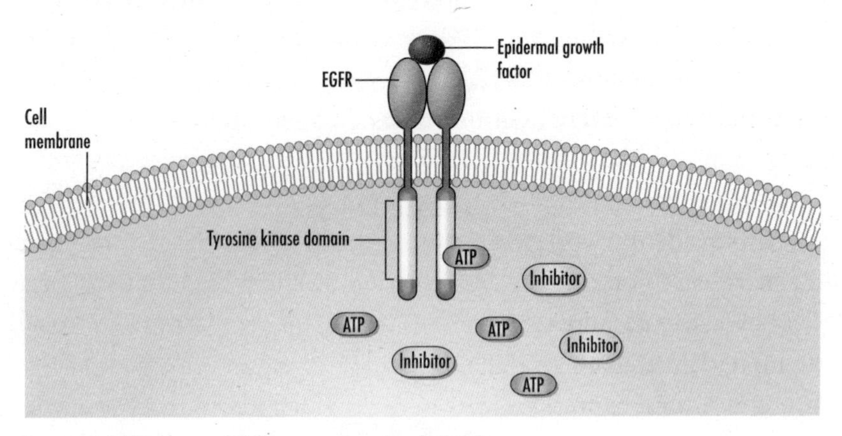

Figure 5. EGFR kinase inhibitors competitively bind to ATP-binding sites in the EGFR tyrosine kinase domain.

* Adenosine triphosphate (ATP) has three phosphate groups, and routinely gives one or two of them up in the course of fundamental metabolic processes. ATP is universally available throughout every living cell, which may be why phosphorylation has evolved to be a primary signalling mechanism.

So why didn't they work in all cancers that overexpressed EGFR?

My colleague Vince Miller catalogued the patient attributes that seemed to correlate with responsiveness to the new drugs: 'We were seeing tumour shrinkage predominantly in people who had never smoked,* people of Asian heritage, women and people with adenocarcinoma histology [a subtype of lung cancer that forms in glandular cells].' Why them?

There were a few potential clues from earlier cancer research. In the 1980s, Ira Pastan at the National Cancer Institute showed that some cancers overexpress EGFR because they harbour extra copies of the *EGFR* gene.[11] Julian Downward and Mike Waterfield of the Imperial Cancer Research Fund showed that the cancer-causing gene in a type of chicken cancer was a *mutated* form of *EGFR*.[12] A decade later, Bert Vogelstein at Johns Hopkins found that the *EGFR* gene was mutated in glioblastoma (a type of brain cancer).[13] In the latter two cases, the mutations in the *EGFR* gene altered the EGFR protein so that it sent excessive signals stimulating perpetual cell proliferation.

Could a mutated *EGFR* gene have something to do with the selective responsiveness of those few patients to the EGFR inhibitors? Or did they harbour extra copies of the *wildtype* (non-mutant) gene?

Alongside my work with patients in the clinic, I also performed basic research on cancers as a postdoctoral fellow in the lab of the Nobel Laureate and CEO of MSKCC, Harold Varmus. Harold's team had created an animal model of human lung cancer in which we could turn expression of a specific oncogene (cancer-causing gene) on or off in a mouse's lungs by adding or taking away a component of the animal's diet.[14] When we turned the gene on, the mouse developed lung tumours over the course of several months. When we turned it off, the tumours melted away within a week. It

* Defined as fewer than one hundred cigarettes in a lifetime.

was like magic! And it conclusively showed that a single mutated gene could cause lung cancer – at least in a mouse.

Seeing disappearing mouse lung tumours in the lab and disappearing human lung tumours in the clinic convinced us there was a genetic basis for the EGFR inhibitors' variable effectiveness. We reasoned that there must be some kind of mutant proteins causing these lung cancers to grow, and the EGFR inhibitors were shutting the proteins off, making the tumours regress. As the drugs were designed to target EGFR, we thought the most likely mutant protein would be EGFR, and so there must be mutations in the *EGFR* gene (or some closely related gene) in these tumours. But we had to prove it.

In the summer of 2003, Harold and I initiated a large-scale (for that time) lung cancer sequencing project with Rick Wilson and Elaine Mardis from the Human Genome Institute at Washington University. We planned to sequence multiple cancer-causing genes from lung cancer tumours, starting with the most likely culprit, *EGFR*. Vince Miller, Mark Kris and others from MSKCC joined in, supplying tumour samples from patients (with their consent).

At that time, it was very difficult to sequence tumour DNA. Normal clinical practice was to embed tumour biopsies in paraffin wax and then cut slices to examine under a microscope. Extracting DNA from paraffin for bulk sequencing was challenging. Eventually, Elaine and Rick's group managed to obtain enough DNA to compare the *EGFR* gene in tumours from patients who responded to gefitinib with the same gene from those who did not.

Harold, Vince and I travelled to St Louis in December 2003 to see the results. The Washington University team had found an *EGFR* mutation in a tumour from a patient who had responded to gefitinib – what we now call an *Exon 19 deletion* (E19del). It was truly a thrilling moment!

But it was only one case. Moreover, it was an unusual mutation, and we didn't know how it might impact the behaviour of the EGFR protein. To be certain it wasn't a fluke result, we had to find at least one more case. Before publishing our results, I also needed to show in biochemical experiments what this mutation was doing to EGFR protein and how the EGFR inhibitors were affecting it.

When you hold a scientific secret like that, you live in fear of getting scooped. In academia, publication is everything, and they who publish first win recognition, grants and prizes. I remember Googling 'EGFR mutation' and 'cancer' regularly through the spring of 2004 to see if anyone else had reported what we had discovered. Meanwhile, I was conducting a series of experiments to show the biological significance of the E19del mutation.

Then, on 15 April, my worst fears were realised.

A Google search revealed that a researcher from another university had reported finding a patient whose lung cancer harboured an *EGFR* mutation. As it turned out, he believed the mutation conferred *resistance* to gefitinib, not *sensitivity* (meaning he thought the mutation made the cancer less, not more, responsive to gefitinib), so my pathway to publication glory might have been secure. Unfortunately, the day soon got a whole lot worse.

Mark Kris called. 'William, are you sitting down?'

'Why?' My heart stopped. I had no idea why he might ask. Was a patient in trouble?

'Because Tom Lynch at Mass General [Massachusetts General Hospital] just sent me a paper that's about to get published in the *New England Journal of Medicine* showing that *EGFR* mutations are associated with sensitivity to gefitinib.'

I was crushed. Then, a few hours later, Harold called. He had been on a visit to Dana-Farber Cancer Institute. 'William, I happened to run into David Livingston [the scientific director], and he said, "Harold, I've got this great story. Come into my office. Our scientists have discovered that *EGFR* mutations are associated with sensitivity to gefitinib. The paper is coming out in *Science*."' I was doubly crushed.

Sometimes, the universe conspires against you. On a single day, I found out that three other research groups had discovered *EGFR* mutations, and two of them understood their role and significance. For lung cancer patients, that day would mark a turning point in clinical best practice. For me, it was one of the worst days of my career.

Science is a competitive sport, and – as in any sport – there are those who win gold and those who don't. Just like athletes, scientists experience the full spectrum of emotions, including rivalry, jealousy and fear of losing out. Ultimately, competition sharpens our focus, makes us do better experiments, and improves the overall level of performance. But that was not how I was feeling at that moment.

After wallowing for a day, we decided we had to get our *EGFR* mutation story out fast, or the other groups would get all the credit and we'd be left with nothing. My wife and I were supposed to go to Hawaii for a long-planned vacation. She said she understood when I cancelled the trip (I still owe her that trip to Hawaii!). For the next year or so, I was in the lab more or less 24/7 to complete the work.

The Dana-Farber and Mass General groups published simultaneously in April; we published in August. We confirmed their results: *EGFR* mutations – including E19del and another mutation called L858R* – were found in tumours that responded to gefitinib but not in those that didn't. We also showed analogous results with erlotinib, indicating for the first time that this drug worked in the same way. Finally, we demonstrated that – among lung cancer patients – never smokers were more likely to have *EGFR* mutations than former or current smokers.† [15]

In the meantime, we had moved on to the next clinical problem.

The new drugs seemed at first to be game changers for patients with *EGFR* mutations. But, to our huge disappointment, most of

* L858R means that the amino acid leucine at position 858 in the protein is changed to arginine. A few other, much rarer, mutations in *EGFR* were also discovered that play a similar role to L858R and E19del.

† The EGFR kinase domain mutations found in lung cancer were different from the ones previously found in glioblastoma and chickens.

these patients started to get sick again after a year. Their tumours returned; the drugs had stopped working. Somehow, the cancer had become resistant to both gefitinib and erlotinib.

Paul Kalanithi's memoir *When Breath Becomes Air* documents this brutal progression.[16] A young neurosurgeon, Kalanithi developed stage-IV EGFR-mutant lung cancer. He was prescribed erlotinib, 'a little white pill', and welcomed the side effect of severe acne that indicated a good response. His health restored, he returned to neurosurgery. Seven months later, a new tumour appeared; the little white pill was no longer working. Chemotherapy failed, and he died aged thirty-seven.[17]

I sat down with Harold to discuss the likely science behind this *acquired resistance*. Oncologists had seen something similar with another kinase inhibitor, imatinib, which seemed to be a miracle treatment for certain forms of leukaemia, but which then stopped working in some patients. Researchers had discovered a newly formed *resistance mutation* in the oncogene that imatinib targeted.

We both agreed this was the most likely explanation: gefitinib and erlotinib were selecting for the emergence of cancer cells containing a *second* mutation in the *EGFR* gene, which in turn was making EGFR protein resistant to the drugs. The cancer was evolving in response to treatment.

Hypothesising such a mutation was easy; proving it, in 2004, was not. We had to show that the mutation – if it existed at all – was present in tumour tissue from patients who had responded to the drugs and then relapsed, but absent in their pre-treatment samples. Since the drugs were still in early development, there was only a small number of such patients. We had to obtain tumour tissue from them, with their consent.

This meant, in most instances, getting second biopsies from patients who were not in the best of health. A lung biopsy is an invasive process that carries risk for the patient: it can cause a dangerous haemorrhage, or even collapse a lung. The risks are worth it for a patient when they first present with suspected lung cancer – the diagnosis is essential to their treatment. But a second biopsy would offer no immediate benefit to the patient.

When Vince and I first proposed re-biopsying patients in early 2004, some of our peers understandably felt it was unethical. However, when we asked patients if they would be willing to undergo a second biopsy, many said yes with enthusiasm. They wanted to understand more about their own disease and potentially help others in the future. Vince and I forged ahead and got the protocol approved. Many patients willingly signed up for our study.

We had reasoned correctly. Of the first six patients re-biopsied, we found a second mutation in three of them.* This new mutation, T790M, had not been detected in the patients' tumours before they started gefitinib or erlotinib; it was selected for during treatment and became more predominant over time as drug resistance developed. We found the same mutation in a human lung cancer cell line which also had the L858R mutation, and we showed that this cell line was resistant to the EGFR inhibitors. Subsequent studies showed that T790M arises in half of patients whose tumours respond to, but then become resistant to, gefitinib or erlotinib.

At last, we had the full genetic picture. AstraZeneca and OSI Pharmaceuticals had intended to develop drugs that would inhibit *wildtype* (non-mutant) EGFR. But in fact the drugs they created, gefitinib and erlotinib, worked best against EGFR that had an *activating mutation* that caused cancer – generally L858R or E19del. These same drugs then drove the evolution of the new T790M mutation – a *resistance mutation* – which curtailed their effectiveness.

Once again, I had a shot at publication glory, and I was determined not to get scooped. As we were finishing up our paper, we heard that our Dana-Farber rivals were working on the same mutation with a team from the Beth Israel Deaconess Medical Center led by Daniel Tenen.[18] This time, we published one day ahead.[19]

* We were able to identify the resistance mutation fairly quickly by searching the analogous part of *EGFR* to the part of the gene that held the imatinib resistance mutation.

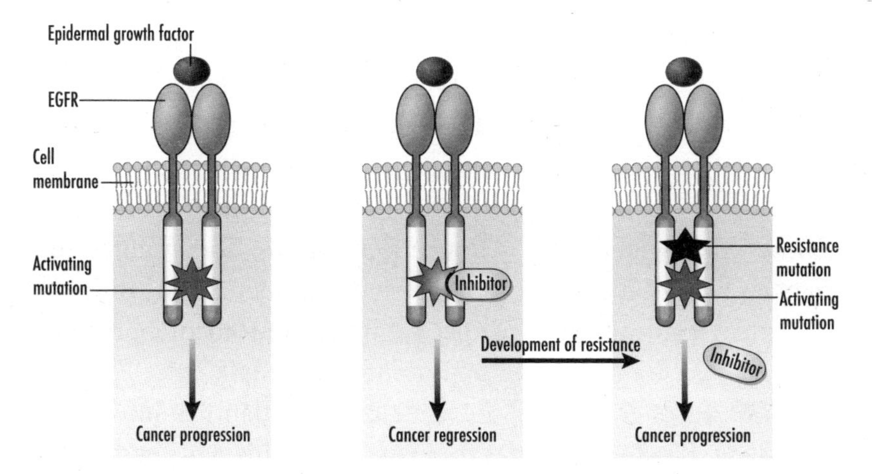

Figure 6. EGFR-mutant tumours are sensitive to EGFR kinase inhibitors, but they develop resistance over time due to the acquisition of second-site mutations in *EGFR*.

More important than publications, of course, was the question of what to do about the second mutation. Neither gefitinib nor erlotinib worked in patients who had developed T790M. My lab developed a treatment regimen combining two other existing EGFR therapies* that had some effect in patients with acquired resistance, but the side effects were often intolerable. New drugs were needed.

Another critical issue was how to convince other oncologists that *EGFR* mutations in lung cancer were even relevant. Looking for mutations in tumours ahead of treatment was unheard of at that time. How could we make it standard practice?

From 2004 onwards, the three hospitals involved in the *EGFR* mutation discoveries – Massachusetts General Hospital, Dana-Farber and Memorial Sloan Kettering Cancer Center – started genotyping patients' tumours routinely, looking for *EGFR* mutations. With test results in hand, our oncologists could tell patients, 'Your tumour has a mutation in a gene that is associated with sensitivity to EGFR inhibitors, so I'm going to prescribe one for you. The side effects will be minimal compared to chemotherapy.'

* Afatinib (a second-generation EGFR inhibitor) and cetuximab.

Conversely, if there was no mutation, we knew to try something else. Gone were the days of blind treatment.

Extraordinarily, influential oncologists at other renowned institutions did not believe our mutation findings. They thought that too many copies of wildtype EGFR was the reason some tumours responded so well to gefitinib and erlotinib, even though the clinical data did not justify their conclusions.

'It's interesting how smart people have trouble getting beyond an idea when the facts say otherwise,' reflects Mark Kris.

It took another five years for mainstream medical opinion to be convinced. Finally, a clinical trial run by AstraZeneca settled the matter: people with lung cancer of known *EGFR*-mutation status were randomised to receive either chemo or gefitinib. Patients with EGFR mutations did better on gefitinib than chemo; patients without mutations did better on chemo.[20] At last, our work was vindicated, but it was extremely frustrating that it took so long.

Before we get to the drug that came out of all this science, let's take a quick look at how the other two research groups found the activating mutations. All three of us achieved this critical discovery in quite different ways, a common occurrence in scientific breakthroughs.

Tom Lynch is a clinician, not a researcher: 'I'm just the doctor,' he says modestly. One of his patients at Mass General was Kate Robbins, a former nurse and mother of two young children, who had stage IV lung cancer. When gefitinib saved her from an imminent death, the story was reported in the *Boston Globe*:

> Last week, Robbins, 46, had her first checkup in five months: Out of 16 once-bulging tumours scattered throughout her organs, there appeared to be none.
>
> 'It's the most amazing thing,' she said.[21]

Daniel Haber read the newspaper story over breakfast and rang Lynch. 'Tom, I think I know why your patient had this incredible response,' he said. 'I think she has a mutation in the tyrosine kinase domain of the EGFR protein.'*

Haber is a leading cancer scientist and physician, also at Mass General, and had just that weekend met with Brian Druker, one of the oncologists who developed imatinib. That got him thinking about kinases (imatinib targets several), and mutations that might render them oncogenic. By an accident of timing, Haber was perfectly primed to see the solution to the scientific riddle posed by a feel-good story in a daily newspaper.

Lynch sent Haber a specimen from one of Robbins' tumours, along with specimens from eight other patients who'd responded to gefitinib. Haber sequenced their *EGFR* genes and found mutations in eight out of the nine – half of them L858R and half of them E19del.[22]

Meanwhile, across town . . .

Pasi Jänne had been treating and researching lung cancer at Dana-Farber since he was a trainee there. His PhD was in Genetics, so he too was well primed to consider a genetic explanation for the strange pattern of response to gefitinib he was seeing in patients.

His mentor, Bruce Johnson, had developed a cell line, H3255, taken from a lung cancer patient in the 1980s. These cancer cells reproduced in a culture that could last indefinitely. Jänne tested gefitinib on a sample of these cells, along with a number of other lung cancer cell lines. The results were remarkable. The H3255 cells were swiftly killed by gefitinib, at very low doses. Jänne found that H3255 was one hundred times more sensitive to the drug than the other lung cancer cell lines. There was clearly something unusual about its genes.

Two other Dana-Farber physician-scientists, Matthew Meyerson and Bill Sellers, were simultaneously investigating the oncogenic potential of tyrosine kinases in lung cancer. A lung cancer genetics specialist, Meyerson knew about the strange results

* Kate Robbins would live almost another two decades.

from the gefitinib trials and suspected there might be a mutation explanation. With Sellers, he analysed the sequences of forty-seven different kinases, including *EGFR*.

Via these two converging routes, the Dana-Farber team identified both the L858R and E19del mutations. They then sequenced the *EGFR* gene from some of Jänne's patients and showed that those patients who benefited from the drug all had one of the mutations; the rest did not.[23]

AstraZeneca had spent much of that decade putting gefitinib through a series of massive cancer trials. They hired more than twenty oncologists from around the world to run the trials, including a young doctor from Romania called Serban Ghiorghiu.

Raised under the old communist regime, Ghiorghiu had dreamed of being an architect. He longed to build something that would last – have an impact. 'In a different world, I might have had other options,' he muses. But in the Ceauşescu era, one of the few attractive professions for a liberal-minded scholar was medicine. As the country reoriented westwards, his oncology professor became involved in clinical trials, and Ghiorghiu joined him as a sub-investigator. The job offer from AstraZeneca came in 2003.

At first, gefitinib was all anyone was talking about. But when multiple negative trial results started to come in, the gefitinib team was cut from over twenty physicians to just Ghiorghiu and his boss. They were left to pick up the pieces and try to figure out why only certain patients responded to the drug.

'The moment when the mutation was discovered, suddenly everything fell into place,' says Ghiorghiu.

After conducting new, patient-selective trials informed by the genetic discoveries, AstraZeneca was eventually able to regain approval for gefitinib – but only for lung cancer patients with the L858R and E19del mutations. To add insult to injury, erlotinib, the rival drug from OSI Pharmaceuticals and their partner Genentech, won FDA approval in 2004 for use in *all* second-line lung cancer patients, not just those with EGFR mutations. (Mutation testing had not been prospectively performed, as it was not standard of care at the time.) The benefit versus placebo for erlotinib was found to be statistically significant – although it only gave patients (on average, both with and without mutations) an extra two months, not the more than twelve months that patients with mutations were getting.* [24]

But bad luck doesn't last forever. A new generation of scientists had joined AstraZeneca while the gefitinib studies were under way, bringing fresh ideas and enthusiasm to the problem of the EGFR resistance mutation. Two young researchers, Darren Cross and Richard Ward, were drawn together at the company's stylish countryside campus at Alderley Park in Cheshire, England.

Cross was a biochemist from Lincolnshire who'd made a significant discovery about an oncogenic kinase while researching his PhD at the University of Dundee;[25] he had pursued his interest in kinases at GSK before being recruited by AstraZeneca in 2001. Ward was a chemist from the south coast of England who specialised in computational chemistry – modelling molecular structure and function on computers to guide drug design; he joined AstraZeneca in 2002, following his PhD. The two men bonded over kinases, and together they pondered what the next wave of kinase inhibitors might look like.

One day, they had a meeting with the head of Chemistry and representatives of AstraZeneca's China business unit. It was 2009, and gefitinib had become an important asset for the China team, due to the higher proportion of lung cancers harbouring EGFR mutations in East Asians. The big problem, of course, was the T790M resistance mutation, which was rendering gefitinib useless for these patients after a year or so.

The China team's request was simple: could Cross and Ward come up with a second-line drug for patients who no longer responded to gefitinib?

* The FDA did eventually limit the use of erlotinib to EGFR-mutant lung cancer in 2016.

Darren Cross had worked on kinase inhibitors targeting various cancers throughout his eight years at AstraZeneca. None of his projects had reached the clinic. 'It gets frustrating – you're on this hamster wheel, working on stuff at full pace, but things don't go forwards, through no fault of your own.' But none of that work was wasted, he says. You always learn something useful. Now, he channelled all that experience into a detailed profile for the new drug.

It must be able to inhibit EGFR with the activating mutations (L858R or E19del) *and* the resistance mutation, T790M – so-called 'double-mutant EGFR'. And where gefitinib hit wildtype (non-mutant) EGFR, causing side effects like severe skin rash and diarrhoea, their ideal drug would selectively target mutant EGFR only. Finally, the drug should be orally bioavailable, to be taken as a tablet ideally once a day.

Meanwhile, Richard Ward was also getting to grips with the T790M mutation. The name comes from the single amino acid change that results from a point mutation in the *EGFR* gene: a **t**hreonine residue is replaced by a **m**ethionine residue at position **790**. Threonine is a small, compact amino acid; methionine is big and bulky. That physical difference completely changes how easily other molecules bind to EGFR.

The threonine residue is located right beside the ATP binding site on EGFR. When the T790M mutation arises, and that threonine changes to a methionine, the larger amino acid makes it harder for gefitinib to fit in the binding site. Curiously, it also makes it *easier* for ATP to bind.* So the competitive balance between gefitinib and ATP shifts. Thanks to the methionine, it's now much more likely that ATP will outcompete gefitinib and slip into the binding site, allowing EGFR to phosphorylate itself and drive cancer.

The first step, Ward decided, was to look for molecules he knew could bind to a kinase with a methionine looming over the

* This critical discovery was made by Michael Eck at Dana-Farber, whose work has also contributed much to the understanding of mutant EGFR. See: https://www.pnas.org/doi/full/10.1073/pnas.0709662105.

binding site, in what is called the *gatekeeper* position. That might sound like a very esoteric search criterion, but in fact Ward had been working with just such a set of compounds. He'd recently come off a project targeting Type 1 insulin-like growth factor receptor (IGF-1R), another tyrosine kinase implicated in cancer. IGF-1R also happens to have a methionine residue in the gatekeeper position.

There is very little of the archetypal chemist about Richard Ward. A keen traveller, sportsman and foodie, his work is conducted on a computer, not a lab bench. Sometimes he makes use of futuristic 3D goggles that allow him to 'see' the shape of a molecule, rotating it to examine all sides, exploring how it might fit into a virtual binding site.

Though he enjoyed chemistry at school, says Ward, 'It was clear early on that I was a pretty dreadful lab chemist.' He never really got to the bottom of this deficiency. 'It's like cooking, I suppose: different people can follow the same recipe and some people get something nice out of it and other people get some black charred thing out of the oven. People in the same lab would get these nice crystals coming out and I would just get this gunge.'

Luckily, a final-year computational chemistry project at the University of Birmingham allowed him to swap his Bunsen burner for a computer, on which he learned to model protein molecules and investigate the fundamental mechanics of their interactions. 'Behind the computational approach is some pretty hardcore physics,' he explains. At last, he'd found a discipline that allowed him to explore chemical compounds without actually having to handle them. He's never looked back.

Ward's earlier work on kinase inhibitors and IGF-1R, coupled with the knowledge and expertise AstraZeneca had built up over many years of research against these targets,[26] meant he did not have to follow the usual pharma path of screening potentially millions of compounds to find a molecule active against double-mutant EGFR. His 'screen' included just forty molecules, of which five were – he already knew – adept at slipping past a methionine gatekeeper in another kinase. He was delighted to find that two of those five showed activity against double-mutant EGFR in *in vitro* assays.

Crucially, the molecules were much less active against wildtype EGFR, which, as mentioned previously, is found in gut and skin cells: they were less likely to cause the debilitating side effects of rash and diarrhoea associated with gefitinib and erlotinib.

For Cross, seeing that *margin* between the effect on the mutant EGFR and the effect on wildtype EGFR in the *in vitro* assays was the eureka moment. 'I still remember standing by the coffee machine when we saw the data, thinking, *yeah, this could be something*.'

Ward and Cross had been operating beneath the radar, running the search for a double-mutant EGFR inhibitor in the downtime from their official assignments. 'It was kind of Black Ops,' laughs Cross. Now, they took these promising early findings to the head of UK oncology discovery, Andrew Mortlock.

A British chemist with a passion for history and nineteenth-century Australian stamps, Mortlock had joined Zeneca at its inception, following his postdoc at UC Berkeley. He's always found chemistry somewhat wondrous: 'It fascinates me that you can make a little white tablet and, depending what's in it, it will burst into flames or cure cancer.' For him, organic chemistry is pleasingly logical, like a language with no irregular forms.

Mortlock had done his fair share of research into kinase inhibitors, and he had a particular interest in how cancers developed resistance to these drugs. So when Cross and Ward turned up with their proposal, he was extremely receptive. He brought in biologist Teresa Klinowska as project lead, and the two pondered how best to structure such a programme.

Ward and Cross were proposing a drug that would be used only when gefitinib and erlotinib no longer worked – a *second-line* medicine to tackle acquired resistance. That significantly limited the number of patients who would be treated: only those lung cancer patients with an EGFR-activating mutation who then developed T790M-based resistance would be eligible for the proposed drug.

Mortlock and Klinowska wanted a drug suitable for more patients. If the drug were designed to work even in the absence of the T790M mutation, then it could be used as a *first-line* drug to rival erlotinib and gefitinib. And if it didn't hit wildtype EGFR, and so had fewer side effects, this new *third-generation* EGFR inhibitor would be *better* for patients than the first-generation inhibitors.* Then the new drug would be the first choice for *any* patient with an EGFR-activating mutation.

'This is not about coming in after gefitinib fails,' they insisted. Mortlock would give the formal go-ahead and allocate the necessary resources for the project on condition they aspired to treat patients in the first-line setting.

For Mortlock, the approval of OSI Pharmaceuticals' erlotinib while gefitinib was withdrawn still stung. 'It was a kick in the teeth,' he says. 'They got it, we didn't.' The new drug could change all that.

Teresa Klinowska grew up surrounded by brainy academics in Cambridge, England, with two scientists for parents. This had its pros and cons. 'It was interesting when I brought my homework down to the dinner table and the assorted range of PhDs couldn't solve it because things had moved on since their time,' she laughs. To forge her own path, she broke away to study biology at the universities of Birmingham and Manchester. Science for her is about the intellectual challenge of solving problems piece by piece, and the excitement of making a real impact for patients with cancer, a disease that has touched too many of her family and friends.

As project lead, Klinowska was in charge of strategy and governance for the third-generation EGFR inhibitor programme. She

* The second-generation EGFR inhibitors included afatinib.

brought in lung cancer experts to work with Cross and Ward, and a medicinal chemistry team headed, at different times, by Michael Waring, Sam Butterworth and Ray Finlay. The medicinal chemists worked with Ward to design new compounds, and then synthesised them in the lab. Meanwhile Cross and the expanded bioscience team developed cell culture assays representing the different EGFR mutations.

From those first two hits – two chemical scaffolds that had the potential to bind to double-mutant EGFR – Ward modelled two series of compounds in consultation with his medicinal chemistry colleagues. The team soon arrived at a set of compounds that were suitably potent against double-mutant EGFR in a test tube. But there was a problem. When Cross and the bioscience team tested the same compounds against *cancer cells* with double-mutant EGFR, they were much less potent. The key difference, they discovered, was the concentration of ATP present in the cells.

You'll recall that EGFR plays a critical role in cancer progression by binding ATP and activating intracellular signalling pathways, and EGFR inhibitors function by blocking that binding. T790M, we learned, makes it easier for ATP to bind, giving it the upper hand over the drugs. The *in vitro* tests had included some ATP in the test tube, but not a lifelike concentration. The cell assays contained much higher concentrations of ATP – as is found in human plasma – and so the ATP was better able to compete with the team's chosen molecules. An inhibitor might bind temporarily to the EGFR binding site, but then it would come loose and an ATP molecule would slip readily in, initiating the cancer signalling pathway all over again.

One way to overcome the ATP competition problem, Ward and Cross decided, was to alter the temporary grip their compounds had on EGFR. Their proposed solution was a compound that would bind *irreversibly*: once it was locked on to EGFR, nothing else would ever be able to attach to that binding site. However easily ATP might bind to double-mutant EGFR, it couldn't outcompete a compound that permanently sealed off the binding site.

This approach had been used before, notably in the second-generation EGFR inhibitor afatinib, which had had significant side effects and had shown disappointing results against double-mutant EGFR tumours.[27] Drug hunters are cautious by nature and prefer to create medicines that can be administered one day and broken down and eliminated from the body the next. An irreversible binder might affect the body for longer, causing significant toxicity problems. So the idea of making a drug with irreversible binding was controversial. 'Irreversible inhibitors had a bad rap,' recalls Mortlock.

Ward believed it should be possible to modify both chemical series to form an irreversible bond with a cysteine residue in the EGFR binding site. But it took a lot of careful thought and risk analysis by Klinowska and the team before they committed to this approach.

'It was at that time, I think, a very brave decision,' says Mortlock.

Ward used computational modelling to adapt the team's current molecules to bind irreversibly. When the lab chemists synthesised the first two compounds, the bioscientists found they showed really strong activity in cellular assays. Irreversible binding had delivered a huge increase in potency. Moreover, the compounds were strongly active against both double-mutant EGFR *and* E19del/L858R-mutant EGFR without the T790M mutation, while sparing the wildtype. Whatever form of EGFR mutation a patient had, these compounds should be effective against it.

Meanwhile, Klinowska and Mortlock were fending off distracting input from other parts of the company. After initial scepticism about the programme, other executives were now offering advice and suggesting alternative approaches. Klinowska politely referred to these interventions as 'challenges of focus'. One group insisted the molecule should target not just EGFR but another cancer-implicated kinase as well. 'It was a raging debate across the industry,' says Klinowska. Some argued that drugs should be like a fine chef's knife – designed to do one thing very well; others believed they should be more like Swiss Army knives, capable of tackling a range of disease causes simultaneously.

Mortlock remembers these debates within AstraZeneca as 'very tense'. He was clear that targeting EGFR alone was challenging enough; expanding the scope would be foolish and would risk all kinds of toxicities. Resisting those calls for extra bells and whistles was, he says, 'one of the critical decisions we got right'. He saw it as his job to protect the team from all this, letting them focus on their original vision. 'One of the things I'm proudest of was being innovative *but not too innovative*. It would have been so incredibly easy to kill the innovation by trying to do too much.'

Susan Galbraith was hired in part to address AstraZeneca's long years of oncology disappointments. A British physician, she was the first R&D head in the company's history who wasn't a chemist. 'They'd been told they were failing,' she recalls. 'Confidence in Oncology R&D was very low. I was well aware that I was coming in as a change agent.'

On arrival in 2010, she axed twenty-five per cent of the portfolio straight away, and redirected resources to the most promising projects. Among these was the third-generation EGFR inhibitor programme. 'It ranked very highly,' she explains. 'We had a clear idea of the patients we wanted to treat; we had a clear idea of the mechanism by which we could improve on the currently available drugs; and we had the chemistry capability to do it.'

Galbraith's parents were both teachers, and she was heavily influenced by a father who had always encouraged his children to ask *Why?* From that simple question stemmed Galbraith's lifelong interest in how the body works. 'I was fascinated by why cancer happens,' she says. Galbraith completed her medical training and a PhD on vascular targeting treatments for cancer. Ward and Cross could not have asked for a better champion of their programme.

Galbraith assigned more chemists to the EGFR inhibitor programme, accelerating the process of synthesising and testing the many hundreds of molecules that Ward and the chemistry team had designed. And when the programme hit its most challenging roadblock, she was ready to make the tough – but necessary – call.

Mark Anderton always suspected there might be an insulin problem with the third-generation molecules. He knew they were derived from kinase inhibitors that Ward had previously identified to target the insulin-like growth factor receptor (IGF-1R) and the structurally related insulin receptor (IR), which is critical to the proper uptake of glucose into cells. If the EGFR inhibitors also blocked these receptors – particularly IR – that could have a detrimental effect on patients' blood sugar levels.

Anderton considers himself a patient and determined person. 'In science, you have to be quite determined because often things won't work in the way you expect,' he says. After growing up in Northamptonshire, his first experience of the pharma industry, through an industrial placement with Novartis in Basel, Switzerland, was an exciting eye-opener. He met his wife there, was awed by the abundant R&D resources the company had at its disposal, and loved being able to ski at the weekend. His PhD at the UK Medical Research Council's Toxicology Unit got him interested in drug safety, and in 2008 he took a job at AstraZeneca as a discovery toxicologist.

This role is now quite common, but at the time it was a novel idea: rather than invent a molecule and then assess whether it is safe or not, why not embed drug safety into the very design of the molecule? 'Let's not wait until we pick the molecule to find it's got problems,' was Anderton's summary.

Joining the EGFR inhibitor programme in 2010, Anderton began thinking through the potential safety risks. As well as the *on-target* risks associated with the inhibition of wildtype EGFR, he was concerned to explore any *off-target* risks, where the drug might hit unintended targets. The most likely were other kinases with similar ATP binding sites. Anderton was worried about the irreversible binding that Ward had engineered into the molecules. If they bound irreversibly to other important kinases, knocking them out permanently, that could lead to safety risks.

He assessed and dismissed this concern relatively quickly. Irreversible binding should only happen if there was a cysteine residue in just the right place in the binding site. Only a few kinases had such a cysteine, and he was able to show they weren't likely to be affected by the team's molecules.

The IR doesn't have a cysteine residue in its binding site, which gave Anderton confidence that any impact on glucose uptake would be short-lived. Nevertheless, he wanted to check it out. He performed a series of studies testing the lead candidate molecule in rats, and saw a worrying spike in blood sugar levels. In the dose range the team believed would be efficacious, the compound was causing hyperglycaemia.

'No one believed me that glucose and insulin were going to be an issue until we performed rat safety studies and said, "There you go!"' he laughs.

The finding led to urgent discussions within the team. Hyperglycaemia – too much sugar in the blood – can be dangerous, but it's a well-understood problem that can be managed. Compared to metastatic lung cancer, it might be considered fairly trivial. Moreover, the related IGF-1R was considered a promising anti-cancer target and it was argued that targeting EGFR and IGF-1R together might increase efficacy. On the other hand, the risk of hyperglycaemia might limit the dose clinicians would be willing to give their patients, which in turn might limit the drug's effectiveness.

It was a really difficult question: was the insulin threat serious enough to undermine the value of the potential drug?

Susan Galbraith remembers this moment as a critical decision point in the programme. They had recently found out that a new biotech firm, Clovis Oncology, had a rival third-generation EGFR inhibitor in the pipeline, and the published data suggested the Clovis drug was further ahead. There was a lot of pressure on the team to deliver fast, to avoid the commercial penalty of trailing behind a competitor. Eliminating the molecule's effect on the insulin receptor – if it was even possible – was likely to take several more months of design and testing, and put them even further behind. It was a tough call, but Galbraith and the team decided the insulin effect was too great a risk to accept.

The chemists went back to the drawing board.

A key part of the molecule was a chlorine group that bound close to the EGFR methionine residue, contributing to potency. This chlorine-methionine interaction really helped with selectivity over wildtype EGFR, which does not have the methionine. But the chlorine was also instrumental in binding the insulin receptor.

So to eliminate the effect on insulin, Richard Ward and lead medicinal chemist Ray Finlay decided to scrap the chlorine group. Potency against double-mutant EGFR slumped, but they were able to modify a different part of the molecule to restore it. Without the chlorine group, the molecule was less selective against the wildtype, but there was still a good margin between mutant and wildtype EGFR impact. With the right dose level, that problem could be managed.

When Anderton repeated the rat studies with the new molecule, the hyperglycaemia problem was gone. The insulin receptor was no longer being hit to the same extent.

As it turned out, the rival drug developed by Clovis Oncology also had an insulin-related problem. Consequently, they encountered significant hold-ups and challenges in their clinical trials. Although Clovis eventually submitted their drug to the regulatory authorities, it failed to get approval due to dose selection and safety challenges, and eventually Clovis terminated the programme.

There was one final potential problem to navigate. When the team tested the revised molecule in mice, they found a portion of it was broken down to form an *active metabolite* – a derivative compound with different pharmacological properties. This is a common stumbling block in drug discovery: if your medicine is broken down by enzymes in the patient's bloodstream into different molecules – metabolites – those new chemical derivatives might have unanticipated and potentially problematic effects before they are excreted.

Anderton tested the active metabolite in rats and was relieved to find it had no effect on the insulin receptor, and there were no other unexpected toxicities. Unfortunately, the metabolite proved to be more active against wildtype EGFR. On the positive side, it was actually more potent against double-mutant EGFR.

This last finding led the team to wonder whether they should transfer their focus to the metabolite – make this new molecule the drug. But its effect on wildtype EGFR soon put to rest that idea.

The critical question was *how much* of the drug would be broken down into the active metabolite? Too much, and they would be back to the bad old side effects of rash and diarrhoea. In mice, the answer was around fifty per cent, but modelling suggested it could be substantially lower in humans. If that turned out to be right, they might avoid those side effects almost entirely.

It was another tough decision for Galbraith and the team. This time, they decided to accept the risk. They would take the molecule, AZD9291, into the clinic, and monitor the proportion broken down into the active metabolite in patients. They hoped it would be low enough to make AZD9291 viable.

I met Darren Cross at a lung cancer conference in California in 2009. He enjoys cracking jokes, but he's also sincerely passionate about his work. He told me about AstraZeneca's third-generation EGFR inhibitor programme and invited my lab to collaborate. Darren sent us multiple compounds, which we tested for efficacy in a blinded fashion against human EGFR-mutant cell lines.

Since early 2004, my lab, together with Katerina Politi in the Varmus Lab, had made transgenic mice that developed human lung cancer tumours with various EGFR mutations (e.g. L858R alone, L858R plus T790M). Among the AstraZeneca compounds we tested in these mouse models was the final lead compound, AZD9291. Sue Ashton was the *in vivo* team lead in Cross's bioscience group.

Together with her team, we demonstrated that AZD9291 was effective against the known EGFR-mutant configurations; older EGFR inhibitors were effective against the activating mutation (L858R) but not the resistant one (T790M).[28]

'That really built our confidence that this molecule should translate into the clinic,' says Cross.

We often encounter unwelcome surprises, like that active metabolite, during drug discovery efforts. But the AstraZeneca team also got one very welcome surprise. Richard Ward, Ray Finlay and the chemistry team had done their best to keep the molecule small and compact, to improve its solubility and pharmacokinetics. This design conferred an unexpected benefit: AZD9291 was able to cross the blood-brain barrier, the selectively permeable border between the blood in the body and the blood in the central nervous system (brain and spinal cord). The blood-brain barrier keeps most pathogens and toxins out of the brain, but it also excludes many pharmaceuticals. This is a problem for cancer drugs if the cancer has spread to the brain.

Cross and Ward did not set out to create a drug that could reach the brain, but serendipity and efficient design had delivered exactly that. A pre-clinical study in an animal model of brain metastases showed good efficacy with AZD9291 where gefitinib had shown very little. If EGFR-mutant cancer metastasised and reached a patient's brain (sadly a common outcome with this type of lung cancer), AZD9291 might have a chance of tackling it.

AstraZeneca hadn't launched a new cancer drug in ten years, so the start of clinical trials was a very big moment for the company. Susan Galbraith asked Serban Ghiorghiu to lead the clinical programme. 'Susan said to me, "You have all that gefitinib experience, you know lung cancer and all the external experts – would you like to take over this new molecule?"'

Galbraith and Ghiorghiu worked with the project team, led by Anne Galer, to design the first study, which would determine the optimum dosage of the drug. Galer had led previous large Phase

III studies, and nobody better understood the importance of achieving robust data to support dose selection. 'Getting the dose right is one of the most fundamental pieces of drug development,' Galbraith explains, 'and it's often the biggest thing that people get wrong.'

As a physician, Galbraith has always enjoyed early human studies more than any other part of the drug discovery process. Initial human responses are the first real proof a drug is going to work. 'You can see a medicine emerging before your eyes,' she says.

One of the first big debates Galbraith and Ghiorghiu had was over patient selection. Should they trial AZD9291 in 'all-comers' (patients whose tumours had wildtype or mutant *EGFR*), or only patients with an EGFR-activating mutation? The latter group was smaller, and it might be difficult to recruit enough patients. But Ghiorghiu had learned the hard way with gefitinib how important patient selection was, and he insisted on enrolling only patients with EGFR-mutant tumours – a decision Galbraith wholeheartedly supported.

Most first-in-human studies take place in the United States or Europe, but to improve their chances of finding enough patients, Ghiorghiu widened this initial study to include Japan and South Korea. EGFR mutations are much more common in East Asia, and thanks to gefitinib, he had built a good network of oncology investigators in the region.

Traditionally, a Phase I cancer study will administer an initial low dose to three patients. Once that has been shown to be safe, three more will be given a slightly higher dose, and so on up through escalating dose levels, until the first signs of toxicity appear. We then drop back down to a lower dose to test for efficacy in Phase II.

Galbraith and Ghiorghiu wanted the first osimertinib study to gather far more detail on the drug's effects at each dose level. So once the first three patients had been dosed at a particular level and had shown no ill effects, investigators would administer the drug to more patients *at that dose*, as well as escalating to the next dose in another three patients. They would treat up to twenty patients at each dose level – a huge number in a first study.

In this way, the team would gather a much richer dataset on the effects of osimertinib at different dose levels. These *expansion cohorts* are now commonly used in drug development, but at the time they were relatively new. The extensive dataset they yielded would be highly persuasive when it came to seeking approval for subsequent studies.

'It made the FDA very comfortable,' says Ghiorghiu, 'it made investigators very comfortable, and it made the Asian countries very comfortable.'

The *endpoint* (indicator of success) agreed with the FDA was *significant tumour shrinkage*. Prior to dosing, patients' lungs were scanned and their tumour diameters measured. They were then measured again six weeks after dosing, and results were confirmed at least four weeks later. The drug would be considered efficacious if both post-treatment CT scans showed average tumour diameter had decreased by at least thirty per cent.[*29]

The study began in March 2013. Of the first six patients given the initial dose of just 20mg – one fiftieth of a gram – three showed a response to osimertinib: their tumours shrank by more than thirty per cent. It was a very promising result at such a low dose.

'I already knew we had a drug at that point,' says Galbraith. 'I was so excited. I knew this was going to be an important medicine.'

'Until it goes into a patient, you're not quite sure if it will deliver the goods,' reflects Cross. 'So when we started to see those first scans coming through, it was really exciting.'

The molecule was incredibly potent. With such encouraging results at low dose levels and an acceptable safety profile, they

* Previously, the FDA had insisted on *progression-free survival time* as the primary metric for cancer drugs, which takes a lot longer to determine, so this was an important innovation for the regulator.

were able to escalate rapidly to 240mg doses. In the end, the team selected a clinical dose of just 80mg. It was, says Ghiorghiu, a landmark decision – a major departure from the bad old chemotherapy days when 'the best dose was the highest [survivable] dose.' He stressed the importance of selecting the optimum dose, 'not only from an efficacy standpoint, but from all different perspectives: safety, brain penetration and patients' quality of life.'

Blood analysis revealed that only ten per cent of the dose was converted into the potentially problematic active metabolite in humans. That was just enough to boost the potency a little, without causing unpleasant toxicity. 'I call it the Goldilocks effect,' says Cross: not too much, not too little metabolite.

I was an investigator on that first study. I saw an excellent response in my patients, and the side effects were very mild – just a little light rash in most cases.[30] Unfortunately, I had to recuse myself when I was offered a job by Roche, a competitor to AstraZeneca. Nevertheless, I followed the published results from afar with mounting excitement. Having shown eight years earlier that the T790M mutation was associated with acquired resistance to EGFR inhibitors, it was extremely gratifying to see a small molecule designed to overcome such resistance actually work in humans. I wished I could have told my father I had made a difference in the lives of cancer patients.

Pasi Jänne also took part as an investigator, treating around thirty patients. He felt a similar thrill at seeing the first positive response in a human to a drug he had long conceptualised. 'It was a grand slam, just an amazing feeling,' he says. 'We saw many, many patients benefit from it.'

Jänne, too, noted the lack of side effects. As every patient in the first study had previously been on either gefitinib or erlotinib, they were in effect their own control arm. 'I asked every single one of them, "How is it being on this drug compared to the prior EGFR inhibitor you were on?" and they always said, "It's night and day different; this is so much better."'

Needless to say, once word of those initial positive responses got out, AstraZeneca had no difficulty at all recruiting EGFR-mutant patients for the remainder of the trial.

In charge of late oncology development in 2013 was Antoine Yver, a French physician who had built a track record of taking experimental drugs through clinical trials in double-quick time.

Yver had spent many years in the clinic before moving across to pharma. His specialty was paediatric cancer, and over the years he had had to witness the deaths of hundreds of children in his care. Each one had left a lasting impression.

His first experience of cancer was as a trainee doctor in Paris in November 1976. He remembers being little more than 'a white coat and stethoscope', with only a year's training and no real understanding of the job. Thrown into a hospital, he was sent by the chief nurse to go and 'check on Marie'.

'I didn't know what "check" meant, but I said, "OK."' When he opened the door to the patient's room, he found a woman of around thirty-five, mother to two kids, with a massive open wound in her chest, bleeding and oozing with pus. She had advanced breast cancer, and the stench of it was terrifying. There was no question that she would be dead soon. In that moment, Yver learned to put aside his scientific training and find the human connection with his patient. Ever since, he says, he has been filled with a sense of urgency to do everything he can for such people.

He was determined to get AZD9291 – now called osimertinib – to patients fast.

Clinical trials generally take more than five years, an achingly slow progression through three phases of study before a medicine can be approved for sale. Yver and Ghiorghiu decided to cut through all that structure and compress the process to its bare essentials. Their overriding philosophy was to figure out exactly what the regulators needed and give it to them as fast as possible.

A typical Phase III study consists of a randomised control trial pitting the experimental drug against placebo or against some comparator regimen. That's the gold standard of evidence. But in discussions with the FDA, Yver and Ghiorghiu came up with a plan whereby AstraZeneca would test for drug efficacy in two *independent* studies, without the need for long and complex randomised control trials. The FDA agreed it would be sufficient to show – in two different studies with two different teams of investigators – that osimertinib shrank tumours significantly in a majority of patients with double-mutant EGFR, with lasting effect.

Meanwhile, Yver and Ghiorghiu pulled every lever they could to accelerate the clinical trial process. They obtained Breakthrough Therapy, Fast Track and Accelerated Approval designations for osimertinib – all FDA initiatives designed to expedite review of potentially important drugs. As EGFR-mutant lung cancer is more common in East Asia, Yver built up AstraZeneca's China R&D team from three people to over three hundred. Through this team, Ghiorghiu worked with Chinese investigators and regulators to set up a key China study that recruited its full complement of double-mutant patients within just seven days.

Part of the recruitment process was the biopsy and genetic test that would determine each patient's EGFR status. Usually, these tests are conducted centrally, adding weeks to the process. AstraZeneca introduced rapid local testing – confirmed later by central testing – to speed enrolment.

A further innovation accelerated patient recruitment even more: new technology to perform liquid biopsies from a blood draw meant investigators no longer needed to wait weeks for a pathology lab somewhere to locate and share the patient's original solid tumour biopsy.

Across the world, 80mg doses were administered to patients with EGFR-mutant lung cancer, with powerfully positive results, even in the brain. In patients with brain metastases, CT scans showed tumour shrinkage in those lesions. Analysis of patients' spinal fluid found concentrations of osimertinib about one fifth of the level in the bloodstream – more than enough for efficacy against mutant EGFR in the brain.

Teresa Klinowska recalls one patient, an AstraZeneca employee who developed brain metastases. She had a young family, and her prospects looked bleak. Osimertinib shrank the tumours in her brain, alleviating her symptoms dramatically and giving her more precious time with her children. When she came to talk to the discovery team about her experience, it really underscored the magnitude of their breakthrough.

The clinical trials did reveal one important safety concern. In 3–4 per cent of patients treated with osimertinib, interstitial lung disease (ILD), an inflammation of the lung, was reported. This condition leads to shortness of breath, eventually reducing the amount of oxygen in the bloodstream. Gefitinib had the same rare side effect, as do many other medicines, including chemotherapeutic agents, antibiotics, antiarrhythmic drugs and immunosuppressive agents. The majority of cases in the osimertinib trial were minor, but a small number of patients died from ILD. This tragic outcome must be weighed, however, against the far greater mortality risk of the cancer: life expectancy with double-mutant lung cancer was just a few months.

Once alerted to the danger, the clinical team were able to define an approach to minimise its severity. They were able to reduce the fatality rate to around one in three hundred patients – a risk heavily outweighed by the benefit of the drug.

By following Ghiorghiu and Yver's accelerated trial design, and completing two independent studies that both showed the same impressive efficacy (with one of them demonstrating long-term durability of response), AstraZeneca was able to submit osimertinib for approval far more quickly than normal. The FDA approved the drug, for use in patients who had been treated with gefitinib or erlotinib and developed resistance, on 13 November 2015, just two years and eight months after the first dose was administered to a human.

Prior to Covid-19, that was the fastest drug development process in history.*

These initial studies assessed osimertinib as a *second-line* therapy, for use only in patients who had become resistant to gefitinib or erlotinib. But Galbraith, Mortlock and Klinowska had long insisted that the drug should be able to do anything those earlier EGFR inhibitors could do, and with milder side effects. They intended osimertinib to be a *first-line* therapy, administered to patients with EGFR mutations as soon as they were diagnosed, and that's how Cross, Ward and the chemistry team had designed and built the drug.

Some in AstraZeneca were worried osimertinib wouldn't perform as well as the first-generation EGFR inhibitors in *treatment-naïve* patients. They felt the company should wait at least until they had second-line approval before investing in the substantial added cost of testing a first-line application. That would require osimertinib to outperform the old rival, erlotinib, in a large randomised trial. Erlotinib at the time was seen as almost unbeatable in patients with EGFR-mutant lung cancer.

But Galbraith, Yver and Ghiorghiu agreed they had to be bold, and in August 2013 AstraZeneca submitted osimertinib to head-to-head trials against the two first-generation therapies.

One of the principal investigators was Jean-Charles Soria, a French oncologist who had worked on the first studies. He championed this new trial and persuaded other investigators to sign up. He encountered a lot of opposition from EGFR specialists. They felt that osimertinib should be reserved as the backup drug for when erlotinib or gefitinib failed. If they gave patients osimertinib straight away, there would be nothing left in the toolbox for when the cancer developed resistance, as it inevitably would.

* Four other drugs were in development with the same double-mutant EGFR target – none of them made it to approval.

But Soria made his case with conviction. 'We were true believers,' he says. 'I spent a lot of time evangelising.' If this *was* objectively the best drug, he and his team argued, patients should have access to it immediately. The early studies showed it could reach metastases in the brain, which the other drugs couldn't do so well, and it had much milder side effects. 'This drug was so good, so much better tolerated than the others.'

Side effects, even severe ones, might not seem like a big deal when one is faced with imminent death, but Soria knew better: 'I never forgot one patient who told me after she was on a first generation EGFR inhibitor, "I will not thank you for the extra year of life you gave me, because my life was miserable."' She was a very beautiful woman, explains Soria, who cared deeply about her appearance. The drug had had a terrible effect on her skin. For her, the extra year wasn't worth a disfigured face.

Soria and his team managed to convince enough oncologists to support the first-line trial, and the results a few years later were clear: osimertinib delivered longer progression-free survival and a longer duration of response in previously untreated EGFR-mutant lung cancer than the other two drugs.[31]

'It was even better than we expected,' says Yver.

Osimertinib was granted approval for use as a first-line treatment by the FDA and other regulatory agencies, and has now almost entirely displaced erlotinib and gefitinib for EGFR-mutant lung cancer therapy. Today, patients treated with osimertinib as a first-line therapy have a median survival of more than three years. With chemotherapy, survival would be less than one year.

We think of innovation usually in terms of spectacular new technologies that allow us to do things we've never done before – streaming movies, generating electricity from sunlight, curing cancer. But some of the most important innovations in human history have simply cut the cost of technology, thus making it broadly available. Similarly, speed can be a valuable dimension of innovation. In both its discovery and development, osimertinib has been remarkable for its accelerated delivery.

Susan Galbraith attributes this speed to the institutional experience gained from gefitinib, the network of lung cancer physicians built up inside and outside the company, and the kinase chemistry that people like Ward and Cross made their passion. Teresa Klinowska emphasised the importance of iteration to the innovation process. 'These things don't come from nowhere. They come from good people having great ideas but working on knowledge accumulated in the past.'

Osimertinib can contain metastatic cancer for some time, but it is not a cure. Generally, within a couple of years, it too is rendered ineffective by emerging resistance in the cancer. 'Tumours are very clever,' sighed Cross. We still don't completely understand the new resistance mechanisms: the cancer is finding other routes to initiate cell proliferation that get around EGFR. When that happens, we usually fall back on chemo.

But, for now, osimertinib remains our first and best defence against EGFR-mutant lung cancer. It has been approved for use in early-stage cancer (before the disease spreads to other tissues) as an adjuvant treatment (to stop cancer returning after surgery). A report at the American Society of Clinical Oncology annual meeting in June 2023 showed that treatment with osimertinib after surgery significantly lowers the risk of death in adults with resected EGFR-mutant lung cancer.[32]

The significance of osimertinib goes beyond its immediate clinical benefit. It was one of the very first drugs deliberately designed to target a specific genetic mutation. Earlier drugs like gefitinib had done so only by chance. The discovery of the EGFR mutations also inspired the development of a lot of other drugs targeting other signalling proteins mutated in cancer.

Lung cancer remains a major killer, but thanks to innovations like osimertinib it's a very different disease than when I started out in oncology. It's now possible for patients with EGFR-mutant lung cancer to live for several years with metastatic disease – and to live well for most of that time.

Osimertinib was approved more than half a century after Rita Levi-Montalcini and Stanley Cohen discovered growth factors. They hadn't intended to develop medicines, and yet the path to a valuable new drug started with their simple question – *What makes cells grow?* It then meandered through new insights, mistakes and misunderstandings, false dawns and endless iteration. Along the way, competition and rivalries in academia and industry spurred discoveries that would deliver a huge impact for patients. The key to it all, though, was that initial desire to figure out how nature works. The quest for fundamental knowledge came far ahead of any practical ambition to create a life-changing medicine.

CHAPTER 3

The Universal Affliction

Condition: Pain
Innovation: Paracetamol [parr-uh-SEE-tuh-mol]
Company: McNeil Laboratories

MOST OF THE MOLECULES IN THIS book are cutting-edge medicines fresh from the laboratory, targeted at terrifying diseases or rare genetic conditions. But for one chapter, let's set aside the X-ray crystallography, gene editing and computational modelling of modern science, and see how drug discovery happened in earlier decades.

'Pain is inevitable,' goes the old saying, 'but suffering is optional.' One of the best means we have to limit the suffering caused by physical pain is chemical modulation of our nervous system. Pain control is still a work in progress, and many of our pharmaceutical solutions come with their own challenges, but we have advanced enormously through the modern era. Some of the most widely used drugs on the planet are painkillers – *analgesics* – that have been known to us for more than a century.

In 1886, the city of Strasbourg was booming. It had been besieged and badly damaged by Prussian bombardment during the Franco-Prussian War, but now, as part of the new Germany forged by Otto von Bismarck, the city was being redeveloped with plentiful investment and a grand building programme. The great Kaiserpalast (Imperial Palace) was under construction, along with boulevards, museums, theatres and a train station. The university had been re-established in 1872, and a splendid observatory was just one manifestation of the city's scientific and cultural renaissance.

Paul Hepp and Arnold Cahn were young physician-scientists working as assistants in the Strasbourg clinic of a celebrated professor of medicine, Adolf Kussmaul. That year, they were assigned to treat a patient with intestinal worms, and they prescribed the standard remedy of the day: naphthalene. You may be familiar with naphthalene from mothballs – it has a very distinctive smell.

The patient took his prescription to the local pharmacy and began treatment. It didn't work: the worms were not eliminated. However, the patient noticed something unexpected. He had been suffering from a fever, and this diminished soon after treatment.

It appeared that the naphthalene had served as an *antipyretic*, a drug that reduces elevated body temperatures. The patient went back to Cahn and Hepp and delivered the surprising news. No one had reported this valuable medicinal property before.

The two doctors consulted the pharmacist, who looked back through his records and realised he had dispensed the wrong compound. Instead of naphthalene, he had actually given their patient *acetanilide*.

The idea that I might prescribe a particular medicine but my patient is given something completely different makes me break out into a cold sweat. But on this occasion, the pharmacist's mistake led to discovery, not catastrophe.

Cahn and Hepp were intrigued. Might acetanilide, never previously used in medicine, be a valuable antipyretic? Fever caused by infections and diseases was uncontrollable in the nineteenth century. For example, malaria was common across Italy and other parts of southern Europe, as well as the southern United States and the tropics. Patients would suffer chills, aches, shivering and fatigue. They would become dehydrated and exhausted, leading to mental confusion, physical collapse and even death. Anything that could help reduce fever would be most beneficial.

Acetanilide is a synthetic substance derived from aniline, which in turn can be made from benzene, the basic ring-shaped molecule that lies at the heart of organic chemistry. During the Industrial Revolution, coal was converted into more useful fuels like coal gas and coke by heating it in an anaerobic environment. A by-product of that process was a thick black liquid, a type of creosote, called *coal tar*. This unpleasant, carcinogenic stuff could be used to preserve timber, seal roads, and – believe it or not – treat skin conditions like psoriasis. In the nineteenth century, it was also the major raw material for the huge and very lucrative synthetic dyes industry.

Benzene, and therefore aniline and acetanilide, were derived from coal tar.

The two physicians responded to their moment of serendipity with cautious optimism – was the observation real and reproducible? They ran a series of experiments, inducing fevers in rabbits and dogs and then treating them with acetanilide. The drug consistently reduced the animals' fevers and did not seem to have any toxic side effects. Encouraged and emboldened, Cahn and Hepp ran a rudimentary clinical trial, treating twenty-four feverish patients with acetanilide. All showed a reduction in fever. No significant side effects were observed, other than a slightly blue colouration in some patients.

That was enough, in the early days of pharmacology, to count as a successful trial. Cahn and Hepp wrote later in the *Centralblatt für Klinische Medizin*, 'A lucky accident dropped a medicine into our hands.'[1]

Paul Hepp's brother worked at a chemical firm named Kalle & Company in Biebrich (now part of the city of Wiesbaden),[2] and he recognised the commercial opportunity acetanilide offered. Kalle ran their own trials, confirming the compound's effectiveness against fever, and decided to mass-produce and market acetanilide as a medicine.

Kalle had no intellectual property claim on acetanilide; anyone could manufacture and sell the stuff. So, in order to conceal their product's true identity, they gave it an alias: 'Antifebrin'. Thus was born, in 1886, one of the first branded medicines. Within a year, the new German wonder drug was in high demand across Europe and America, especially once it was found to have analgesic as well as antipyretic properties – it reduced pain as well as fever. Not knowing what was in it, pharmacists had to order the Kalle branded product rather than purchase acetanilide more cheaply from local manufacturers.

Kalle is still going. Its main products these days are sausage casings and sponge cloths, but at the time it was known primarily as a paint company.[3] Antifebrin seems to have been one of its few forays into the pharmaceutical business. The English-language label on a 1906 tin suggests using the drug for 'nervous affections, facial neuralgia, locomotor ataxia, sciatica [. . .] intensive reflex headaches [. . .] and in headaches of various sorts'. Patients and physicians were instructed to dissolve the drug in boiling water and then allow to cool before ingesting. The label helpfully added, 'If Whisky or Brandy is used as a solvent, NO HEAT is needed.'[4]

However, there was a problem. Blue skin colouration recurred in an increasing number of patients as Antifebrin was more widely dispensed. Blue skin is the manifestation of cyanosis, as this effect is known, and usually indicates a reduction in oxygen bound to haemoglobin. It's an important warning sign of low oxygen levels in the blood. Acetanilide seemed to be having a detrimental impact on haemoglobin.

Concern about the possible toxicity of Antifebrin was raised as early as 1888 by a physician in Comanche, Texas, following the death of a young woman who had been dosing herself regularly to treat headaches.[5] Aniline, from which acetanilide is derived, was known to be toxic to the blood,[6] although thanks to Kalle's secrecy regarding their formulation, most people were unaware of the close relationship between Antifebrin and the poisonous coal tar derivative. We now know that both substances convert haemoglobin into a related molecule, methaemoglobin, which, due to its chemical structure, cannot bind oxygen.

Today, if a medicine was found to be interfering with the blood's ability to carry oxygen, it would be swiftly banned by the FDA* and other regulators around the world. At the end of the nineteenth century, however, no such governmental powers existed, and Antifebrin remained on sale to the public for decades. Instead, it was up to the market to try to find a better, safer analgesic.

The market, in this case, was helped by another big dose of serendipity.

In the years before the pharmaceutical industry got established, many chemists found employment in the invention and manufacture of dyes for the textile industry. One of the first synthetic dyes to be developed was mauveine, which, like acetanilide, is derived from aniline.† Originally known as aniline purple, mauveine was discovered in 1856, and it led to a boom in aniline dye research across Europe. Numerous companies were established to capitalise on the synthetic dye goldrush, among them Bayer A.G. of Barmen (now part of the city of Wuppertal) in Germany.

Strasbourg, Wiesbaden and Wuppertal – the axis of nineteenth-century aniline chemistry – are linked by the river Rhine, the economic backbone of industrial Europe. Even today, most of the continent's major pharmaceutical companies are located on the Rhine, including Roche in Basel, where I worked for eight years.

Bayer was founded in 1863, and initially manufactured aniline and a magenta dye called fuchsine. As the company grew, expanding its dye portfolio, it began accumulating large volumes of apparently useless by-products. One of these was paranitrophenol.[7] Wondering what to do with their stockpiles of this white crystalline material, Bayer's chemists tried modifying it in various ways.

* Founded in 1906.

† Mauveine was invented by mistake by an eighteen-year-old student, William Henry Perkin, who was trying to synthesise a drug, quinine. The dyestuffs and pharma industries have a curiously interdependent history.

One of the chemists, Oscar Hinsberg, added an acetyl group (CH_3CO) to paranitrophenol in 1887, creating *4-epoxy acetanilide*. Aware of Cahn and Hepp's discovery the previous year,[8] Hinsberg decided to test this close relative of acetanilide to see if it too had analgesic and antipyretic properties.

A rather charming description of the nascent clinical trial process is captured in an American patent dispute case of 1901:

> Hinsberg produced and discovered the new body, and suggested its utility in medicine, but as he was a chemist, and not a physician, he was unable to sufficiently test its effects as a medicine by submitting his new drugs to clinical experiments and observations, and therefore associated himself with Prof. Kast for these purposes. Of course, after the discovery of a medicinal body, it is necessary that its action should be tried for a long period to determine the best way of administering it, and in what cases it should and should not be administered, and what are its effects, both ultimate and approximate, before it can safely be placed upon the general market.[9]

Hinsberg's clinical partner was Alfred Kast of the University of Freiburg, and he was able to deliver very good news for Bayer. Not only did this new substance – the chance adaptation of an unwanted by-product in a dyestuffs company – turn out to offer excellent pain relief and fever reduction, it also seemed to be less toxic than Antifebrin, with none of the risks to haemoglobin and oxygen circulation associated with the Kalle medicine.

Carl Duisberg, a young dyestuffs chemist who would go on to become Bayer's legendary leader and develop the foundations of the modern pharmaceutical industry, recognised the great commercial potential in Hinsberg's discovery, and quickly built up the manufacturing capacity to produce it at scale. Bayer may have come newly and unintentionally to the pharmaceutical business, but they were quick learners. Acknowledging the success of Kalle's 'Antifebrin' brand name, they devised a catchy brand of their own for 4-epoxy acetanilide: 'Phenacetin'. Duisberg would later describe the discovery of Phenacetin as 'pure luck'.[10]

In a little over a year, the emergence of Antifebrin and Phenacetin had launched a whole new class of drugs: the 'coal tar analgesics'.

Phenacetin was quickly embraced by the medical profession and patients alike. This may have been helped by the fact that the drug was clearly fun to take: one 1887 paper noted that most users became 'euphoric'.[11] They felt more relaxed, became more talkative and enjoyed an increased appetite. Taking Phenacetin must have been like knocking back a few cocktails.

In the United States alone, between 1890 and 1900, Phenacetin sales were estimated at 150 million doses.[12] The drug's success spurred Bayer to create a whole new pharmaceuticals division, which would – only a few years later – deliver another blockbuster analgesic, perhaps the most iconic of all drugs: Aspirin.

Although generally considered safer than Antifebrin, Phenacetin did nevertheless turn out to have some toxicity issues. Kidney damage was the most worrying, with some heavy users of the drug suffering gradual renal failure.[13] And some patients' skin developed a bluish tinge. Still, Phenacetin remained on sale for almost a hundred years, often in combination with Aspirin as a dual-action pain reliever. It was an extremely successful product in its time. The end came when Phenacetin was linked to certain forms of cancer, especially upper urinary tract cancer. It was banned in most countries by the 1980s.

So, what could be found that was safer than Antifebrin and Phenacetin? Curiously, for the scientists at Bayer the answer was hiding in plain sight right from the beginning. To manufacture Phenacetin from paranitrophenol, they had performed a series of chemical reactions, producing a number of intermediate compounds. One of those intermediate compounds was N-acetyl p-aminophenol.

Today, it is better known as paracetamol.

This substance, first synthesised in 1878 by Harmon Northrop Morse of Johns Hopkins University, was little understood in those early years of pharmacology. Bayer wondered if it might have some medicinal value, and in 1893 invited Joseph von Mering, a renowned physiologist, to assess it. Von Mering was the first person to discover (with Oskar Minkowski) that removal of the pancreas causes diabetes. He also helped to discover barbiturates. His trials found N-acetyl p-aminophenol to be an excellent analgesic and antipyretic. But he declared it to have the same blood toxicity problems as acetanilide – the tendency to modify haemoglobin such that it will not bind oxygen.

This finding, we now know, was entirely incorrect, and may have resulted from a contaminated drug sample. Nevertheless, such was von Mering's reputation that no one ventured to develop paracetamol as an analgesic, despite a number of other studies showing its effectiveness. The wonder drug that hundreds of millions of people use regularly today was shelved for decades on the basis of a single flawed tox report.

Indeed, we might never have rediscovered the virtues of paracetamol if it weren't for a vital insight into the mode of action of our two old aniline-derivative friends, Antifebrin and Phenacetin.

The first hint came in 1895. One of Hinsberg's colleagues, a chemist by the name of Treupel, analysed the urine of patients treated with Phenacetin and concluded that the analgesic was being converted back into paracetamol in the bloodstream.[14] But because Phenacetin was – at that time – considered a safe, effective analgesic, while paracetamol supposedly had toxicity concerns, this analysis led nowhere. The development of Aspirin soon afterwards helped stifle any commercial urge to investigate the compound that was, it seemed, both a raw material for, and metabolite of, Phenacetin.

By 1946, the New York City Public Health Department was growing concerned about the side effects of Antifebrin and Phenacetin.

Both drugs had been widely dispensed across America for decades, and the methaemoglobin problem and anaemia related to methaemoglobin (methaemoglobinaemia) was cropping up again and again among heavy users. It seemed that both painkillers were converting haemoglobin into methaemoglobin and so hindering oxygen transport around the body. The head of the Health Department's Laboratory of Industrial Hygiene turned to one of his lab technicians, a man named Julius Axelrod, to investigate the health risks.[15]

Julie, as he was generally known, was the son of Polish immigrants. The family had no money for his education, but Axelrod was able to attend the City College of New York, where tuition was free, majoring in biology and chemistry. He wanted to study medicine but was refused a place at all the schools he applied to, possibly because quotas limiting the number of Jewish medical students had been introduced a few years earlier.[16] Instead, faced with the global depression of the 1930s, he took a $25/month lab technician job at New York University and for thirteen years made do with a support role in the medical-scientific firmament.

Axelrod's primary responsibility at the Laboratory of Industrial Hygiene, where he moved in 1935, was to measure vitamin concentrations in food samples. While working there, he was involved in a lab accident and lost an eye. It must have been a traumatic experience, but it had the benefit of keeping him out of the war after Japan bombed Pearl Harbor.

'Successful scientists,' Axelrod later wrote, 'are generally recognised at a young age. They go to the best schools on scholarships, receive their postdoctoral training fellowships at prestigious laboratories, and publish early. None of this happened to me.'[17] He was stoic about the cards he had been dealt: 'I expected that I would remain in the Laboratory of Industrial Hygiene for the rest of my working life.'

The invitation to work on Antifebrin and Phenacetin in 1946 changed everything, setting Axelrod on a path that would lead to numerous breakthroughs and even a Nobel Prize.

Axelrod had little experience in this kind of research. On the recommendation of his boss, he reached out to a more established scientist, Bernard Brodie. A British pharmacologist, Brodie had developed his scientific career in the United States, working first at New York University before moving to Goldwater Memorial Hospital* on what is now Roosevelt Island in the East River.

They met in February 1946. 'It was a fateful meeting for me,' wrote Axelrod. 'Talking to Brodie about research was one of my most stimulating experiences.' After years of hard grind as a lab technician, Axelrod had finally found a first-class scientific mind to partner with him and mentor him. The two men discussed how they might try to discover what was causing methaemoglobinaemia in heavy users of Antifebrin (acetanilide). Was the chemical itself having this detrimental effect? Or was it acetanilide being broken down in the bloodstream into an active metabolite (see Chapter 2) that was doing the damage?

A likely culprit was aniline itself, already known to cause methaemoglobinaemia. So Axelrod's first move was to find out if acetanilide broke down in the body to produce aniline.

At Brodie's invitation, Axelrod moved into the lab at Goldwater Memorial Hospital. Guided by the more experienced scientist, he developed a new method for measuring tiny concentrations of aniline. That done, he administered acetanilide to volunteers and tested their blood and urine. Sure enough, aniline was found in both. Moreover, Axelrod was able to show a direct correlation between the level of aniline in the blood and the amount of methaemoglobin present.

'This was my first taste of real research, and I loved it,' he exulted.[18]

Very little acetanilide was found in the urine, implying that it was almost all being metabolised – turned into something else in the body. But what? To answer that question, Axelrod had to develop further tests to detect other possible metabolites. His

* The hospital was demolished in 2014.

results are captured in his first ever scientific paper, written with Brodie and published in 1948:

> N-acetyl p-aminophenol, which is found in both plasma and urine after the administration of acetanilide, appears to be the first step in the major route of the metabolism of the drug. The high concentration of this metabolite in the plasma prompted an appraisal of its analgesic effect. Studies conducted on human subjects [. . .] showed N-acetyl p-aminophenol to be, dose for dose, about equal in analgesic activity to acetanilide.[19]

N-acetyl p-aminophenol, you will remember, is better known as paracetamol.

With Brodie's help, Axelrod had just shown that Antifebrin – the widely used but problematic coal tar analgesic – was broken down in the blood to form paracetamol, and that paracetamol was itself just as good an analgesic as Antifebrin. Those two facts together strongly suggested that it was paracetamol, *not* acetanilide, that was doing the painkilling work. Moreover, paracetamol, when administered in its pure form to volunteers, did not result in significant amounts of methaemoglobin.

Paracetamol, Axelrod realised, was doing the work of Antifebrin and yet was much safer. The lab technician who had been refused entry to medical school and had never before written a scientific paper now held in his hand the key to a new and much better painkiller.

This was a huge step forward in pharmacological thinking. During the nineteenth century, chemical compounds were tested for their medicinal benefits and then simply adopted wholesale by patients and physicians, warts and all. Now, researchers were starting to examine what actually happened to a compound inside the body, and they were realising that a greater understanding of the life cycle of that compound could lead to the development of a better drug. Fundamental knowledge was driving innovation.

As often seems to happen, Axelrod and Brodie were not the only researchers turning their critical gaze on Antifebrin in 1946.

David Lester and Leon Greenberg performed a similar analysis of the blood and urine of patients treated with acetanilide, and they too identified paracetamol as one of the main metabolites.[20] By treating rats with large doses of paracetamol, they too showed that it did not cause methaemoglobinaemia. Lester and Greenberg published their results in 1947, beating Axelrod and Brodie to the glory of the acetanilide revelation.

But Axelrod wasn't done yet. He turned next to the other great coal tar analgesic, Phenacetin.

When administered in high doses to dogs, Phenacetin had also been shown to cause methaemoglobinaemia. Again, Axelrod needed to develop new, specialised tests to measure tiny concentrations of Phenacetin and its possible metabolites. With these tests, he and Brodie were able to show that the methaemoglobinaemia was being caused by a Phenacetin metabolite called p-phenetidine.

But their greatest revelation was the other metabolite they found. Once again, most of the drug was rapidly being converted in the bloodstream into paracetamol.[21]

Just as with Antifebrin, Axelrod and Brodie had discovered that the compound actually doing the beneficial work in people treated with Phenacetin was paracetamol – and, crucially, it was not responsible for the deleterious side effects. In both cases, the products people bought were being broken down by the body to create the real medicine.

So why not just give people paracetamol instead?

The timing of paracetamol's rebirth was perfect. Clinicians were increasingly raising doubts about the safety of Aspirin, the household favourite.[22] And recently, an American gastroenterologist, James Roth, had shown that Aspirin could inhibit blood clotting and cause bleeding in the stomach and upper intestine.[23] A safer alternative was needed.

Yet the pharmaceutical giants of the age did not seize the patent-free opportunity laid at their feet by Axelrod and Brodie. Perhaps they were too invested in their existing Aspirin-based products to risk cannibalising their sales with a new upstart painkiller.[24] Instead, it fell to a small American firm, McNeil Laboratories, to launch the first paracetamol-based medicine.

The company had started life as a Philadelphia drugstore with a soda fountain, and had moved into drug research and production in the 1920s.[25] Robert McNeil Jr, grandson of the founder and by then research director, was already looking for an alternative analgesic. With the help of James Roth, he ran clinical trials that found paracetamol to be as effective as Aspirin in fever reduction and pain relief, and showed it did not have the same gastrointestinal bleeding safety concerns.[26] Convinced of paracetamol's medical value, McNeil coined the generic name most commonly used in the United States, acetaminophen, and began commercialising the drug.[27]

The first product, Algoson, launched in 1953, combined paracetamol with sodium butabarbital, a barbiturate used to treat anxiety and aid sleep.[28] This was followed two years later by a single-ingredient paracetamol medication aimed at children. To brand it, McNeil Laboratories selected letters from the longer scientific name for the compound, *N-ace***TYL** *p-aminoph***ENOL**.[29]

Tylenol Elixir for Children was sold on prescription in a box shaped like a fire engine, with the slogan, 'for little hotheads'.

McNeil Laboratories was acquired by Johnson & Johnson in 1959. The pharma giant proceeded to market the drug to adults and make Tylenol the global brand it still is today.*

Neither Axelrod nor Brodie ever profited from their paracetamol discoveries, but they both went on to pursue illustrious careers at the US National Institutes of Health. Indeed, Axelrod's work on neurotransmitter hormones – for which he won a Nobel Prize – laid the groundwork for the development of antidepressants such as Prozac.[30] 'F. Scott Fitzgerald once stated that there are no second acts in American lives,' he wrote later. 'After a mediocre first act, my second act was a smash.'[31]

* * *

* In the UK, paracetamol was launched as Panadol by Sterling-Winthrop.

Paracetamol has its drawbacks, like any drug. Taken in very large quantities, it can cause liver damage. This is most commonly seen in people who have attempted suicide by deliberately overdosing. But, like Aspirin, its side effects are well tolerated by most people who use it sensibly, and it remains one of our most widely dispensed and generally useful analgesics.

It's remarkable that for more than fifty years, paracetamol was known but ignored due to what was considered scientific dogma from a luminary in the field (von Mering). There's an important lesson here for innovators: we must have the courage and open-mindedness to challenge received wisdom and question even giants in the field, particularly when new data suggest that past conclusions are wrong.

CHAPTER 4

The Royal Disease

Disease: Haemophilia A
Molecule: Factor VIII
Innovation: Emicizumab [em-ee-SI-zoo-mab]
Company: Chugai Pharmaceuticals

WHEN PRINCESS ALIX OF HESSE GAVE birth to a son, her husband, Tsar Nicholas II, was overjoyed. Their first four children had all been girls, and after nearly a decade of trying for an heir, the royal couple had grown increasingly desperate. The Russian people considered Alix – or Alexandra Feodorovna as she was known to them – bad luck. Diplomats across Europe reported the intense disappointment each new girl brought, and Queen Victoria, grandmother to Alix, wrote expressing regret.

So the relief when Alexei Nikolaevich was born in 1904 was not just personal – it was political. In this baby boy, all the hopes and expectations of the Russian Empire were vested.

He was a large baby and seemed strong. But there was an early sign that all was not well: when the umbilical cord was cut, Alexei bled for hours from his navel. Despite copious bandages and the close attention of royal physicians, his blood simply would not clot.

The condition, not properly understood at the time, was nevertheless frighteningly familiar to Alix and her extended family. Her uncle, Prince Leopold, Duke of Albany, had bled profusely from time to time, and had suffered severe joint pain. At the age of just thirty, he had slipped and hit his head, dying shortly after. Her brother, Prince Friedrich, cut his ear at the age of two and bled for three days. And just a few months before Alexei was born, Alix's four-year-old nephew, Prince Heinrich of Prussia, had hit his head and died. He and his brother both had the same tendency to bleed uncontrollably.

This 'royal disease', as it became known, was haemophilia, a genetic disorder that afflicts mostly males. It was first named by Friedrich Hopff and Johann Lukas Schönlein of Zurich University, who came up with the term 'haemorrhaphilia', shortened to 'haemophilia' in 1828. The word comes from Greek roots *haima* (blood) and *philia* (love, or tendency towards).

We know now the genes associated with haemophilia are located on the X chromosome. In humans, the X and Y chromosomes determine the biological sex of an individual: females (XX) inherit an X chromosome from the father, while males (XY) inherit a Y chromosome from the father. Mothers only pass on X chromosomes. Haemophilia is a recessive condition, meaning that women who carry a gene for haemophilia on only one of their X chromosomes do not develop the disorder. Since men only have one X chromosome, they will get haemophilia if they inherit the mutated gene. A woman can develop haemophilia if her father has the disease and her mother is a carrier; this happens more rarely.

We now believe that Queen Victoria was a carrier of a haemophilia gene, which she passed on to her son, Leopold, her granddaughter Alix, and her great-grandson Alexei.

None of this was known at the time. But the Tsar and Tsarina understood that their newborn son, expected one day to rule a vast empire that stretched from Finland to the Bering Sea, was dangerously vulnerable. A moderate blow to the head could kill him. Prince Leopold had been barred from serving in the British army, or taking part in any vigorous activities, yet a simple fall had been enough to cause a brain haemorrhage (bleeding). Alexei would have to be even more carefully protected.

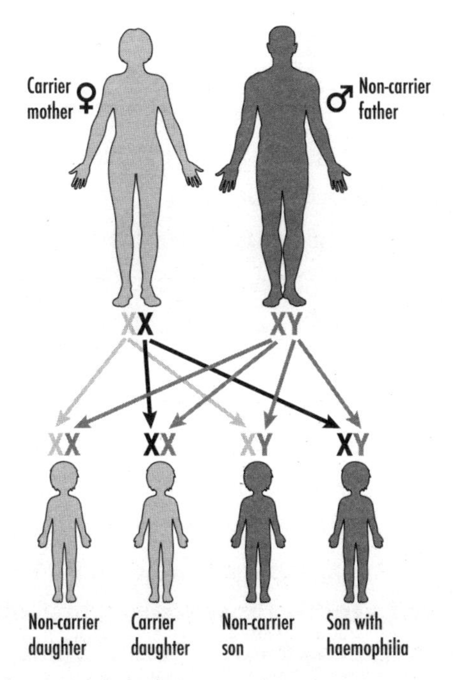

Figure 7. Haemophilia is an X-linked recessive disease.

This is one of the toughest things about living with haemophilia for the more than one million men who have the condition today. The young Alexei was forbidden to ride a bicycle, play tennis or mount a horse. His playtime with other boys was closely supervised to ensure he was never at risk of being knocked over. Even today, people with severe haemophilia are unable to pursue physically demanding careers or play high risk sports, for fear of suffering a fatal haemorrhage. The Tsarevich could not understand why his sisters were allowed to enjoy these activities while he was kept indoors, and it made him restless and resentful.

Internal bleeding can be very painful, particularly in the joints, and Alexei suffered from prolonged haemorrhages caused by simple bruises and falls. Internal bleeding can occur even when sitting still; people with severe haemophilia can get spontaneous bleeds at any time. Alexei's limbs would swell, he would be unable to bend or unbend his arms and legs, and he would scream from the pain. At these times, he couldn't walk. Massage helped alleviate the pain, but also risked further bleeding.

When internal bleeding occurs regularly in joints, the tissue starts to break down, leading to arthritis or even the complete destruction of the joint. People with severe haemophilia sometimes end up in wheelchairs for the rest of their lives. In Alexei's case, however, the brutal progression of his disease was halted prematurely: the Bolsheviks murdered him and his family in 1918.

Almost a century later, at the other end of Asia, Kunihiro Hattori was pondering blood. Unlike the Tsarina, who wanted to encourage clotting to stop her son's abnormal bleeding, he was trying to work out how to *prevent* clotting. Clotting can be as dangerous as excessive bleeding if it occurs inside the blood vessels: strokes, heart attacks, deep venous thrombosis and pulmonary embolisms all result from clots blocking critical blood flow. Hattori wanted to develop a drug that could help prevent such thrombotic events.

A molecular biologist by training, Hattori is a determined, dedicated scientist with a keen focus on social justice. He worked for the pharmaceutical company Chugai in Gotemba City, Japan, at the foot of snow-capped Mount Fuji. Chugai, once known for making vitamin D, was just starting to do interesting work with monoclonal antibodies.

It was the year 2000, and monoclonal antibodies were one of the most exciting new technologies in drug discovery. An *antibody* is a key component of the immune system: it's a protein designed to recognise a specific *antigen*, which is usually some kind of foreign body.* This might be, for example, a virus, a bacterium or a poison. An antibody binds to its specific antigen and either disables it directly or causes other components of the immune system to contain or eliminate it.

* Not all antigens are foreign – people with autoimmune disorders develop antibodies against certain parts of their own body.

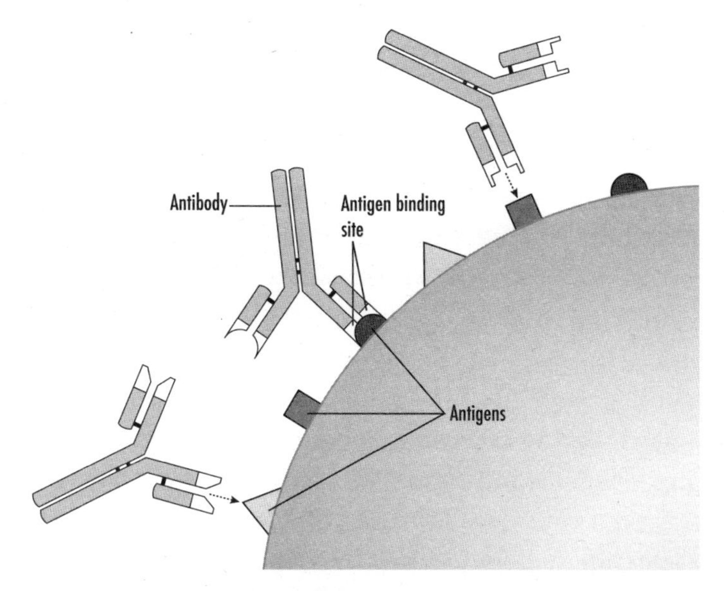

Figure 8. The mode of action of antibodies.

This binding specificity made antibodies particularly interesting to drug hunters, as it offered a way to hit precise disease targets in the body. If an antibody could be built in the lab to target, for example, a molecule on the surface of cancer cells, it might be possible to make a medicine that would locate and eliminate only cancerous cells.

Inhibition of a target antigen was exactly what Hattori was trying to do with the anti-thrombosis programme. He wanted to create an antibody against *Tissue Factor*, a protein that stimulates blood to clot. Such an antibody, he theorised, might lessen clotting and so reduce the likelihood of heart attacks in people with a history of thrombosis, hypertension or diabetes.

Meanwhile, Hattori's Coagulation Research Group was exploring additional research avenues. During a brainstorming meeting, one of his colleagues suggested a possible idea to improve haemophilia treatment.

Hattori hadn't previously considered haemophilia an interesting subject to pursue. Treatment options had improved considerably since the days of the last Tsarevich, and many men – at least in developed countries – were now able to live safely with haemophilia. But Hattori's colleague explained the considerable challenges that people with haemophilia still faced: they needed intravenous injections several times a week, and some developed an immune response against their own treatment.

Hattori listened carefully to his younger colleague's pitch. He ultimately dismissed the proposed drug concept as unworkable, but it got him thinking. Was there a better way of treating haemophilia?

At the dawn of the twentieth century, when Princess Alix was searching for some kind of remedy for her son, heavy bleeds could only be treated with large infusions of another person's blood. But in the 1920s, physicians discovered that plasma (blood without the blood cells) from another person served to stem bleeding. Something in the plasma was helping to initiate coagulation. Unfortunately, a lot of plasma was needed to have any effect, and in most parts of the world it was simply not available.

One researcher who took an early interest in haemophilia was Robert Gwyn Macfarlane, a young English doctor with a passion for ice skating and speeding around on motorcycles.[1] Possessed of thick black eyebrows and a charming smile, Macfarlane has been described as 'delightful, modest and an almost self-effacing figure'.[2] As a medical student, he was assigned to dress a wound on a patient's chin. It would not stop bleeding. The patient, he was told, came from a family of 'bleeders': two of his brothers had bitten their tongues and subsequently died.

Macfarlane recalled this formative experience in a later speech:

> He was a young man, with a small cut on his chin, which bled persistently. We tried dressings and bandages and strapping, the blood just went on oozing through everything. The cut was sewn up, with the result that he got an enormous haematoma, and the stitches burst. Then it was cauterised, but after a day or two a large slough separated and he was bleeding from a wound much

> larger than the original cut. We tried adrenaline, ice packs, hot packs, pressure dressings and things got steadily worse. Meanwhile he had lost so much blood that he had to be transfused, quite a major proceeding in those days.[3]

After the blood transfusion, the bleeding stopped. Something in the alien blood had finally made clotting possible. This fascinating outcome set Macfarlane on a lifelong quest to understand how blood worked.

While still a student, he wrote 'Haemophilia – a short survey', describing the condition in detail and concluding that the only effective treatment was a transfusion of blood or plasma. Aware that haemophilia predominantly afflicted men, he then pursued a rather eccentric research lead: to 'feminise' men with the condition using female hormones. 'It had been claimed that ovarian extracts or ovarian grafting had had dramatic results,' he later recalled. 'I became deeply involved with virgin mice and pregnant mares, though the details are mercifully hazy.'

A more promising approach was suggested by the physiologist Hamilton Hartridge: he had noted that certain snake venoms cause blood to clot. Macfarlane visited London Zoo and introduced himself to the Curator of Reptiles, Burgess Barnett, who enthusiastically suggested they 'milk' all the venomous snakes and test their venoms on haemophilic blood *in vitro*. Macfarlane said afterwards, 'I developed a real aversion to these animals and often had nightmares about them.'

Of the roughly twenty venoms they tested, some were anticoagulants while others had only marginal clotting effect. But one stood out: the venom of Russell's viper (*Daboia russelii*), at a dilution of one part in ten thousand, could clot haemophilic blood in just seventeen seconds. Macfarlane and Barnett tried using the diluted venom to treat the wounds of haemophilia patients and it seemed to work. They even used it for one patient who needed a tooth extraction, successfully plugging the hole with diluted venom. Following their 1934 paper in *The Lancet*, the pharmaceutical company Burroughs Wellcome marketed the venom as a treatment for haemophilia under the brand name Stypven.

Macfarlane went on to spend most of his career as a clinical pathologist at the Radcliffe Infirmary in Oxford, where he established a world-famous haematology department. During the 1950s he led a team investigating blood clotting, and – in partnership with another renowned haematologist, Rosemary Biggs – was able to map the various key molecules, or *factors*, essential to the process.

In 1964, Macfarlane published a landmark paper in *Nature*: 'An Enzyme Cascade in the Blood Clotting Mechanism, and its Function as a Biochemical Amplifier'.[4] As is so often the case, others in the haematology field made similar discoveries around the same time.* At last the mystery of coagulation had been deciphered.

Blood is a miraculous substance. It must flow smoothly as a liquid through our arteries and veins, carrying oxygen, nutrients, waste products, chemical signals, immunity agents and much more around our bodies. Yet should any vessel rupture, it can swiftly turn solid, forming a clot, to slow and stop the loss of blood.

This physiological miracle is achieved through a complex set of molecular interactions we now call the coagulation cascade. A blood clot is composed of platelets – small, colourless cell fragments – held together in a tangled net of fibrin, a fibrous protein. The coagulation cascade is a sequence of molecular interactions that leads to the formation of fibrin. Each step sees one molecule catalyse the formation of another from its inactive precursor.

Working backwards from the endpoint, solid strands of fibrin are formed from dissolved fibrinogen in the presence of the enzyme thrombin. Thrombin is formed from prothrombin in the presence

* Earl Davie and Oscar Ratnoff published 'Waterfall Sequence for Intrinsic Blood Clotting' in *Science* in the same year. See: https://www.science.org/doi/10.1126/science.145.3638.1310.

of a molecule called Factor Xa.* Factor Xa is formed from Factor X in the presence of Factor IXa and Factor VIIIa.† Factor IXa is the activated form of Factor IX, while Factor VIIIa is the activated form of Factor VIII.

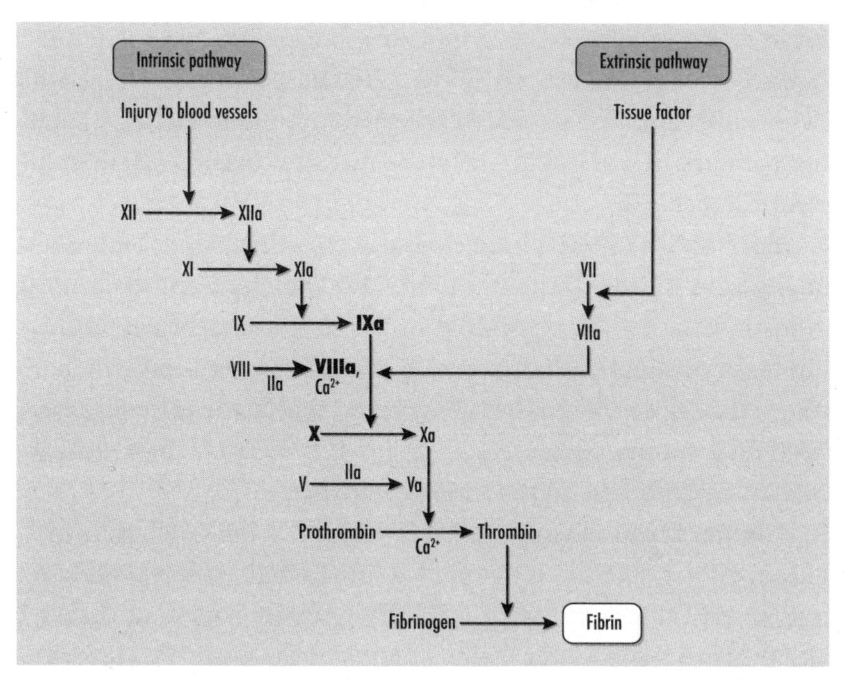

Figure 9. The coagulation cascade.

There's more to it than that, but we don't need to go any further to understand haemophilia. People with haemophilia A (the most common form) are unable to produce enough Factor VIII; people with haemophilia B lack Factor IX. When one of these coagulation factors is missing, the coagulation cascade grinds to a halt and very little fibrin is produced. Without sufficient fibrin, blood can't clot and major injury to a blood vessel can be fatal.

So what can we use to treat people with haemophilia? The first and most obvious answer was *replacement factor*: Factor VIII for

* *X* is the Roman numeral for 10; *a* stands for 'activated'.

† Russell's viper venom, it turns out, directly activates Factor X, making the earlier steps in the cascade redundant.

people with haemophilia A; Factor IX for those with haemophilia B. We'll be focusing on haemophilia A and Factor VIII.

The problem was how to get enough of the stuff into the patient. In the early days, as mentioned above, people with haemophilia were treated with infusions of whole blood or plasma. But this achieved only low levels of coagulation because the infused plasma (with Factor VIII) was so diluted by the patient's own plasma (without Factor VIII). Even if enough donor plasma was available, the patient's body couldn't take the quantity needed to get sufficient Factor VIII.

Macfarlane's initial solution was cows: bovine blood is far richer in Factor VIII than human blood, and the large domestic meat industry meant it was plentifully available. His team began producing bovine Factor VIII from ox blood and treated patients with some success. However, they soon found that many patients developed an immune response against bovine Factor VIII, which the human body recognised as a foreign molecule.

A better solution came in 1965, when a Stanford physiologist, Judith Graham Pool, developed a methodology for producing concentrated human Factor VIII. By freezing and then thawing blood plasma, she found it was possible to produce a *cryoprecipitate* rich in Factor VIII. This could be refrozen and stored in a relatively low-cost process that was soon rolled out to blood banks around the world. By the 1970s, people with haemophilia A had a practical treatment they could reconstitute and self-administer at home as soon as a bleed started, or even prophylactically ahead of physical activities. Life expectancy doubled to around forty-two years.[5] 'It was a great, miraculous gift to the world,' wrote haematologist Carol Kasper. 'Many tens of thousands of patients have benefitted.'[6]

But Pool's great gift to the world had an unintended and tragic consequence. The vast amount of human plasma needed to derive sufficient cryoprecipitate – and the next generation *glycine-precipitated plasma fractions* – came from tens of thousands of blood donors, mostly in the United States. Unfortunately, in the

1970s and 1980s, deadly new infectious diseases – acquired immunodeficiency syndrome (AIDS) and hepatitis C – started to spread through the American population, and supposedly healthy donors inadvertently made haemophilia patients sick.

Hepatitis C and human immunodeficiency virus (HIV), the cause of AIDS, could survive the extraction process and live on in Factor VIII concentrates. Blood donors at that time weren't screened for these viruses. Almost half of all Americans with haemophilia became infected with HIV, and most of those died of AIDS [See Chapter 8].[7] Others became infected with hepatitis C. There were similar catastrophes in other countries. Horrified physicians and patients stopped using replacement factor except for the most serious bleeds, a treatment regression that inevitably led to further deaths.

A desperate scramble ensued to cleanse Factor VIII concentrates, and by 1985 a combination of heat treatment and solvent detergent extraction eliminated the transmission of HIV in the treatment of haemophilia. Further efforts yielded increasingly pure Factor VIII. However, the ongoing challenges of securing enough Factor VIII from blood donations, and the shock of the HIV contamination disaster, galvanised researchers to find an alternative for people with haemophilia A.

These were the early days of biotech, and scientists were beginning to identify the genes that encoded individual proteins. Once a full-length gene sequence was known, the door was open to synthesising its protein in the lab.

To sequence a gene, researchers first needed to obtain a very pure sample of the corresponding protein, and prior to the 1980s no one had ever isolated pure Factor VIII. Two research groups achieved this milestone in 1981–82, paving the way for a commercial race between two recently founded biotech firms to synthesise Factor VIII. In one lane was San Francisco's Genentech; in the other was Boston's Genetics Institute. Genentech had managed to create synthetic insulin in 1978. Genetics Institute, founded in 1980, made Factor VIII one of its first targets.

By 1984, both companies had sequenced the gene for Factor VIII, using completely different approaches.[8] It was by far the largest gene yet identified. Insulin, the great cloning success of the previous decade, consists of twenty-one amino acids. By contrast, Factor VIII has 2,332. Mapping the whole sequence was an extraordinary achievement, and it laid a foundation of experience and technologies that turbocharged the coming biotech revolution.

The size and complexity of the Factor VIII gene threw up further challenges in synthesising the protein. Genentech had previously grown insulin in bacteria, but this didn't work for Factor VIII; instead, they developed a technique to grow the molecule in living cells (Chinese hamster ovary cells). Another eight years would pass before a recombinant Factor VIII product was approved for use in haemophilia. Recombinate from Genetics Institute (partnering with Baxter International) was licensed in the United States and Canada in 1992.[9] Kogenate from Genentech* was approved the following year.[10]

The new synthetic Factor VIII worked just as well as plasma-derived Factor VIII, and it avoided even the theoretical possibility of viral contamination. At last, people with haemophilia A had a truly safe and effective treatment.

Yet while the safety and efficacy of haemophilia A treatment had improved enormously, the therapeutic regimen was still challenging. Replacement Factor VIII is a protein, and so cannot be absorbed from the intestine without being digested: it cannot be swallowed in a convenient pill. It can't even be absorbed via subcutaneous or intramuscular injection, but must be injected directly into a blood vessel.

Worse, the half-life of Factor VIII is only around twelve hours: within half a day of injecting it, half the dose is degraded. So people with severe haemophilia A typically had to inject themselves in a vein every two or three days to maintain a minimal level of clotting

* Licensed to a subsidiary of Bayer.

potential. That's bad enough for an adult; for a child with haemophilia, it is miserable. Extended-half-life versions of Factor VIII have now been developed, but users still need to inject themselves once or twice a week.

A further complication comes from the body's own immune system. In around twenty per cent of people with haemophilia A, replacement Factor VIII provokes an immune response in the form of antibodies called *inhibitors*. These prevent replacement factor from working properly. Treatment for people with inhibitors is extremely complex and costly. Often a *bypassing agent* is used, which is designed to activate a later stage of the coagulation cascade. Bypassing agents must be administered carefully, and can cause life-threatening clots.

Midori Shima is a senior paediatrician at Nara Medical University. The university hospital is Japan's primary treatment centre for haemophilia. During more than forty years of medical practice, Shima has treated many patients with the condition, and his experiences underscore just how challenging it can be.

One of his patients is a software engineer we'll call Kenji. He has severe haemophilia A and his body makes high concentrations of inhibitors against replacement Factor VIII. Throughout his life, Kenji has had to exercise extreme caution. He chose a job that required minimal physical activity. He avoided hot baths, which he feared might increase the risk of bleeding. At night, he would try not to move in his sleep, to avoid causing an internal bleed.

For all his caution, Kenji was unlucky. As a young man, he broke his femur and needed surgery and internal fixation with a metal plate. Shima was with him throughout the operation, administering bypassing agents to try to help his blood clot. His careful monitoring and medication were successful, and Kenji made a full recovery. Nevertheless, he continued to suffer painful bleeding in his joints and muscles. One muscle bleed was so bad he had to be hospitalised for two weeks.

Just one day after leaving the hospital, Kenji fell off his bicycle and broke six bones.

This would be a complex medical problem for anyone. For Kenji, it was a life-threatening disaster. The surgery was successful but, due to the haemophilia, it took a long time for his wounds to heal, leaving him vulnerable to infection. A large abscess developed, and Kenji contracted sepsis. The infection worsened, and eventually the surgeons had to amputate Kenji's whole leg at the hip joint.

The amputation left a large open wound that required long-term haemostatic treatment. It took several months to fully control the bleeding, with regular transfusions to replace all the lost blood.

That is why Kunihiro Hattori was interested in haemophilia A: despite the brilliant work of haematologists and biotech firms throughout the twentieth century, there remained considerable unmet need which translated into serious trauma for people like Kenji. Replacement Factor VIII just wasn't good enough to protect people with haemophilia A with certainty, or allow them a decent quality of life – especially if they had inhibitors.

So what could be used instead of Factor VIII?

Factor VIIIa (the activated form of Factor VIII) works by helping an enzyme, Factor IXa, to catalyse the activation of Factor X. We call molecules that help enzymes like this *cofactors*. Factor VIIIa binds to the other two molecules and attaches to the surface of a cell, enabling Factor X activation and the progression of the coagulation cascade.

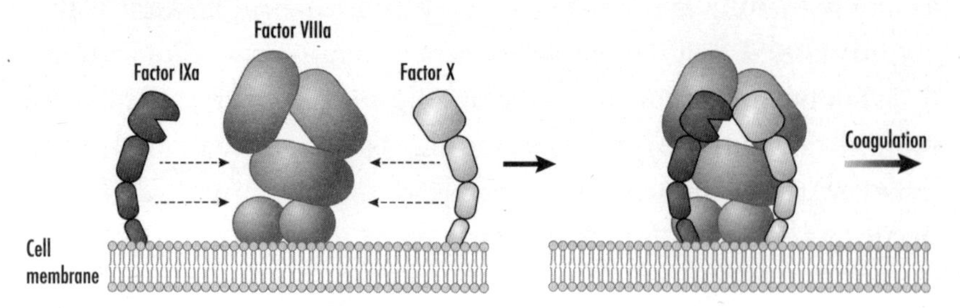

Figure 10. Factor VIIIa is a cofactor that promotes the interaction between Factor IXa and Factor X.

Contemplating this mechanism, Hattori recalled another cofactor he had heard about at a symposium a few years before. That cofactor worked by fixing the catalytic centre of its enzyme at a specific height above the cell surface to make it precisely accessible to the molecule it would act upon. We might liken it to a clamp that holds a test tube just the right height above a flame. The critical scaffold role of the cofactor lies in physically positioning the other two molecules in just the right places to interact with each other.

'Perhaps,' he mused, 'Factor VIIIa might work the same way.'

No one knew for sure how Factor VIIIa did its job, but Hattori's hypothesis laid the foundations for his second big conceptual breakthrough. What could he use as a clamp to hold Factor IXa and Factor X in just the right position above the cell surface to catalyse activation?

The answer he came up with was something never seen in nature.

A bispecific antibody.

It was an extraordinary cognitive leap. The primary purpose of natural antibodies, as high school biology students know, is to detect antigens and trigger the immune system to inhibit or eliminate them or the pathogens that express them. By contrast, Hattori wanted to use an antibody to link two proteins together. While other researchers were inventing new antibodies to eliminate stuff, Hattori hoped to discover one that could create a harmonious union.

It might have advantages over replacement Factor VIII, he reasoned. Antibodies last for weeks in the bloodstream, and can be administered subcutaneously, eliminating the need for regular intravenous injections. Moreover, as a completely different type of molecule, an antibody would not be affected by the inhibitors that rendered replacement Factor VIII ineffective in patients like Kenji.*

An *FVIII function-mimetic antibody* (an antibody that mimics the function of Factor VIII) would be a potential game-changer for

* Although it is always possible that other anti-drug antibodies could emerge.

people with haemophilia A. But a regular antibody would not be sufficient: antibodies function by binding extremely well to a single target, whether it be a virus, a cancer cell or a toxin. That's what every antibody in every animal throughout history has always done. One antibody, one target.

Hattori needed his antibody to bind two different targets at the same time.

Fortunately – and this is a theme we see repeatedly in innovation – an unrelated discovery would give Hattori the technology he needed: the *bispecific antibody*.

What exactly is an antibody? There are various types, but the most common is Immunoglobulin G (IgG), which consists of four polypeptides bound together (polypeptides are chains of amino acids). Two of the IgG polypeptides are identical shorter chains; two are identical longer chains. These are called the light chains and the heavy chains, respectively.

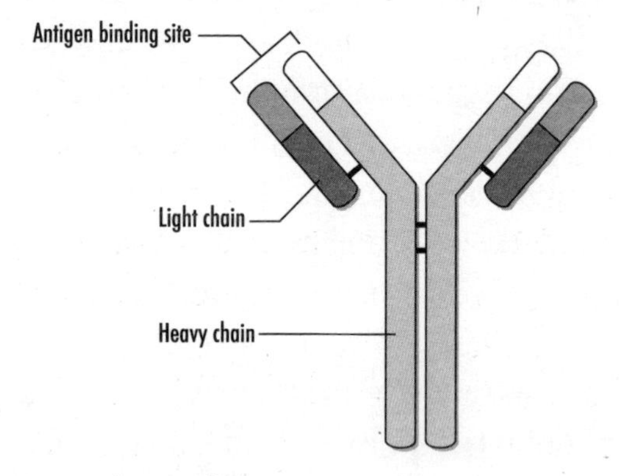

Figure 11. The structure of an antibody.

The two light chains and two heavy chains are arranged in a Y-shape, and the upper branches of the Y are the binding sites that bind to an antigen. Because the light chain and heavy chain on each side of the Y are identical, the binding sites are also identical: each upper branch of the Y will bind to the same type of antigen.

So an IgG antibody is a Y-shaped structure, with a binding site on each upper branch, and both binding sites bind to the same antigen.

Perhaps it was inevitable that some molecular biologist with an engineering mindset would eventually look at that Y shape and ask themselves: *what if each branch could bind to a different antigen?*

The first researcher to consider the idea was American chemist Alfred Nisonoff, writing in 1960 before the Y shape was fully understood. All he knew at the time was that breaking down an antibody with a digestive enzyme yielded two identical antigen-binding fragments.[11] 'It should also be of interest to attempt to prepare antibody of mixed specificity,' he speculated.[12] The following year, he was able to combine two fragments from different antibodies, creating a chimeric molecule that he later showed did indeed bind two different targets.

Back then, however, nobody could do much with this conceptual breakthrough, for one simple reason. We didn't have any practical way to manufacture antibodies.

It's straightforward to make an antibody to a particular antigen *inside a body*. You simply provoke the immune system into producing your desired antibody by injecting the antigen into the bloodstream – that is the basis of vaccination. It's much harder to make an antibody *outside a body*, as we would have to do if we wanted to manufacture an antibody-based medicine.

Early researchers made some progress in the lab. They would inject a mouse with the relevant antigen, stimulating its spleen to create B lymphocytes, a type of white blood cell. Those lymphocytes produced the antibodies specific to that antigen. They could then extract some of the spleen cells that made those lymphocytes, and for a short while these produced the desired antibody in a petri dish. But what researchers couldn't do in the 1960s was grow those spleen cells in a self-perpetuating culture that would endlessly produce antibody. Outside the body, the spleen cells would soon die.

Then, in 1975, German biologist Georges J.F. Köhler and Argentinian biochemist César Milstein discovered an ingenious way to keep antibody-producing cells alive indefinitely, for which they were awarded a Nobel Prize.[13] They fused those fragile antibody-producing spleen cells with cancerous myeloma cells, which have the ability to replicate indefinitely. The result was a *hybridoma*, an antibody-producing cell that is effectively immortal. Because the hybridoma reliably reproduces itself in a petri dish, we can harvest identical antibodies ('monoclonal antibodies') from the culture for as long as we like. Grow the culture big enough and we have an industrial antibody manufacturing process, with the hybridoma cells acting as the factories, churning out antibodies.

Monoclonal antibodies now form a huge and very exciting field of pharmaceutical research. Scientists are designing novel antibodies to fight cancer, pathogens and autoimmune diseases, as well as for use in diagnostics and even pregnancy tests. The vast majority are *monospecific* – each branch of the Y-shaped molecule binds the same antigen.

So what about *bispecific* monoclonal antibodies?

The concept lay dormant until the 1980s, when a flurry of new interest arose. César Milstein created a hybrid hybridoma in 1983, dubbed a *quadroma*.[14] This cell, bred from two different hybridoma parent cells, could produce two different types of light chain and two different types of heavy chain that would then combine to form a bispecific antibody. This was a huge step forward from Nisonoff's technique of using enzymes to tear antibodies apart and then reassembling the fragments. That blunt approach inevitably would damage the delicate protein chains. By contrast, Milstein's quadroma allowed him to synthesise and assemble the component parts of a bispecific antibody all inside a cell.

Over the next few years, researchers developed other techniques to make bispecific antibodies. At the same time, exciting therapeutic applications were conceived. In 1985, Michael Bevan and colleagues showed how a bispecific antibody could be used to link a cancer cell to a 'killer' T cell, focusing the immune system's destructive power against the cancer.[15] This sparked a flurry of work which has recently led to the approval of a handful of bispecific 'T cell engaging' antibodies for use against different types of cancer.

However, by this point a fundamental problem with the technology was becoming evident. It was getting easier to design a bispecific antibody, and even produce small quantities, but the yield was far too low for any commercial manufacturing operation to be viable.

It all came down to the *chain association problem*.

Remember how Milstein invented the quadroma, the hybrid cell that could synthesise two types of light chains and two types of heavy chains? The cell would then assemble these to form the bispecific antibody. But it could also assemble them in a different way, to form an unwanted molecule. In fact, this is what happened most of the time. The cell could assemble a molecule with two of the same type of light chain. Or a molecule with two of the same type of heavy chain. Or with the different chains in the wrong places. Just one of the ten different molecules formed in the quadroma would actually be the correct bispecific antibody. All the others would be wasteful impurities that had to be separated out from the final drug.

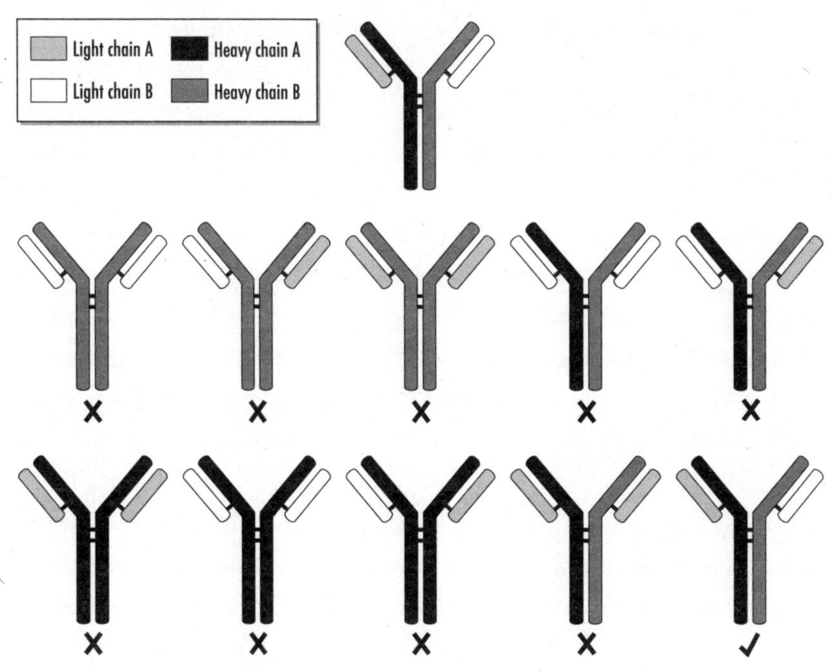

Figure 12. The multiple ways that two types of heavy chains and two types of light chains can assemble.

It's already time-consuming and expensive to make drugs by growing them in cells ('large molecules' or 'biologics'), as opposed to combining chemicals ('small molecules'). It's much, much worse when only one eighth of the product is of any use.* Add to that the serious technical challenge of separating out the unwanted molecules, and you can see why the difficulties of actually making a medicine this way rapidly overwhelmed the early promise of bispecific antibodies.

This is a reality of innovation that is not widely understood. Many great ideas never make it to fruition because the functioning prototype can't be reproduced at scale efficiently enough to be cost-effective.

By the end of the twentieth century, the shine had come off bispecific antibodies. Clinical trial results were disappointing. Significant toxicity issues arose. The pharmaceutical industry was unwilling to fund further expensive trials. And would-be manufacturers of bispecific antibodies were pulling out, unable to solve the chain association problem and improve on the very low yields it caused.[16]

After fifteen years of intense research and development, not a single bispecific antibody had been approved for use as a medicine. While monospecific antibody programmes surged ahead, research labs quietly shelved their bispecific cousins.

That was precisely the moment Hattori decided to create a bispecific antibody to mimic Factor VIII. He wanted to design an antibody to do something diametrically opposed to normal antibody function, using an unproven technology that was becoming increasingly discredited. It was a big, bold leap into the dark.

Kunihiro Hattori is approaching retirement. He has a placid round face and a businesslike manner that speaks of a serious career built of long hours in the lab. These days, he enjoys birdwatching

* There are sixteen different ways to assemble the ten combinations illustrated. Two of them are the desired antibody. See: https://doi.org/10.4161/mabs.21379.

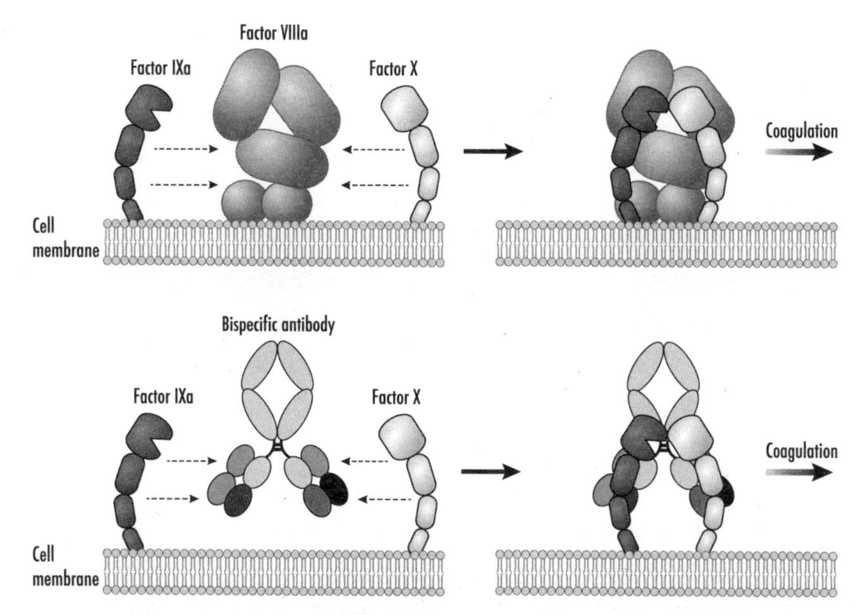

Figure 13. A bispecific antibody could mimic Factor VIII activity by binding to Factors IXa and X and promoting the interaction between them on a cell membrane.

in his neighbourhood park, but back in 2000 he had no time for hobbies. If he wasn't working, he was drinking with his fellow researchers. 'Anything!' he declared. 'Beer, sake, whisky – anything containing alcohol.'

From the beginning, his ambition to treat haemophilia with a bispecific antibody was met by his superiors with extreme scepticism. Hattori had hoped to start work in 2001, but didn't have enough researchers to get going. Most of his own team were still busy on the Anti-Tissue Factor programme, and senior management refused to allocate more people to his haemophilia dream.

For two years, he was stuck.

In 2002, the Anti-Tissue Factor programme was terminated. Disappointing as that was, it meant Hattori could reassign some of the staff to the haemophilia project. Then they had to move very fast.

Antibodies are made in the lab by injecting antigens into mice or other mammals. But Chugai was just about to close its animal facility for several months of fumigation. First staff shortages, now facilities – Hattori feared his bispecific antibody research might be delayed for ever.

One of the young biologists on his team was Takehisa Kitazawa, a bright-eyed researcher full of energy and enthusiasm. Hearing about the fumigation delay, he went to see Hattori and said, 'Let's just start – right now.'

A generation younger than his boss, Kitazawa has a cheerful, more casual manner, although he remains appropriately deferential. He was a track and field athlete at the time – a discus thrower. During lunch breaks at the Gotemba City R&D site, he would head outdoors and hurl a metal disc around Chugai's spacious grounds. The appeal of the discus was straightforward for him: 'It flies very beautifully,' he says. Glinting in the sunlight below Mount Fuji, it must have been a marvellous stress-reliever.

In those few remaining days before the animal facility was closed down, Kitazawa made the first attempt to create a bispecific antibody. He injected one set of mice with Factor IXa and another set with Factor X, the two molecules that Factor VIIIa must bind to and link together. In this way, he provoked the mice's immune systems to create antibodies to the two coagulation factors. Taking spleen cells from each mouse and fusing them with myeloma cells, following Milstein and Köhler's methodology, the team created hybridomas from which they could clone antibodies.

'We made twenty Factor IXa antibodies and twenty-three Factor X antibodies,' recalls Hiroyuki Saito, an immunologist who, like Kitazawa, had worked for Hattori on the terminated Anti-Tissue Factor programme. They then recombinantly created 460 (20x23) bispecific antibodies.

To test whether any of them could mimic the function of Factor VIII, the team created *in vitro* assays consisting of purified coagulation factors. The assays were sensitive enough to detect even slight activity. Ten of the 460 bispecific antibodies were found to accelerate the activation of Factor X in the presence of Factor IXa. It was an early, thrilling, score. Next, they tested those ten antibodies on haemophilia A plasma. Like magic, the experimental antibodies sped up clotting – the higher the dose, the greater the effect.

The very idea that one could artificially fuse two antibodies to make something not found in nature, and that this chimera would somehow do the highly specific linking job of Factor VIII, was wildly optimistic. Yet here, already, was proof that Hattori's bold idea really could work.

'That was very exciting – too exciting!' says Kitazawa.

The team was delighted. It's rare to get such a quick positive result in drug discovery, particularly when pursuing a completely new concept. Now, Hattori recognised, they needed some expert help. None of them had worked on haemophilia before. They needed to talk to someone who understood the disease more deeply.

Hattori consulted a globally prominent researcher in the thrombosis and haemostasis field, Koji Suzuki, who had collaborated on Chugai's anti-thrombosis programmes. Suzuki knew exactly whom to recommend: 'You need to speak to Professor Yoshioka at Nara Medical University'. There and then, Suzuki phoned Yoshioka. After listening to Suzuki's brief explanation, Yoshioka said, 'Please tell them to come to Nara.'

Hattori knew Professor Akira Yoshioka by reputation, through the Japanese Society on Thrombosis and Haemostasis, but they had never previously spoken. At Yoshioka's invitation, Hattori and Saito travelled to Nara Medical University a few days later. It was a hot August day. Hattori presented the concept and experimental data to Yoshioka and his colleague, Associate Professor Midori Shima. 'Chugai needs help evaluating these antibodies,' he concluded.

Shima later recalled his initial puzzlement in that meeting: Chugai had shown no previous interest in haemophilia, yet suddenly they had a solution? Shima had never heard of bispecific antibodies. Like Yoshioka, he was a leading expert on Factor VIII, with numerous papers to his name, but he had not really considered its bridging mechanism – the mechanical way it connects Factor IXa and Factor X – and so was surprised by the mimicking model proposed. Nevertheless he was impressed by Hattori. 'He was very quiet but very strong. I felt he was a real scientist, not a company man.'

Professor Yoshioka listened carefully to the Chugai presentation and reviewed all the data. He made some suggestions for how they might evaluate the bispecific antibodies. After a long, intense discussion, he came to a decision. 'We would be glad to collaborate with you,' he declared.

'They were the top haemophilia experts in Japan,' says Hattori, 'and they agreed this could be a potential game-changer. I was delighted.'

So began a partnership that would last more than a decade.

Professor Yoshioka and Associate Professor Shima were both researchers as well as clinicians, and they helped refine a series of assays that the Chugai team could use to test their candidate molecules. It wasn't enough to determine whether blood clotted – most will eventually. Hattori and his colleagues wanted to know the *speed* of clotting. Yoshioka's team applied an existing technology, clot waveform analysis, to haemophilia – something that had never been done before. This is an often overlooked type of innovation – finding a valuable new use for an older invention. The process involves passing a light beam through a sample of blood plasma and measuring the change in absorbance of light that occurs as fibrin is formed. This delivers a very precise, real-time measure of the clotting rate. The Nara innovation neatly demonstrated the effectiveness of Chugai's bispecific antibodies.

Meanwhile, the next generation effort at the Mount Fuji lab was under way. It was much more ambitious. Molecular scientist Tetsuo Kojima constructed a more efficient experimental system that could generate a much larger number of bispecific antibodies. Kojima and his colleagues, Taro Miyazaki and Tetsuhiro Soeda, successfully prepared as many as four thousand different bispecific antibodies. From this number, the team was able to pick out stronger candidates that led to faster clotting. They made some small modifications to improve their potency, and soon had a molecule that really did speed up clotting dramatically.

But there was another important challenge to overcome. At the back of his mind, throughout the programme, Hattori had known they would have to address the chain association problem that had always bedevilled the viability of bispecific antibody medicines. Now was the time.

Tom Igawa is a chemical engineer. He's also a very enthusiastic cyclist. He's taken his bike all over the world, voyaging across alien lands with nothing but a tent and a sleeping bag. He's explored the Andes and Iran, and cycled from Pakistan to China, right over the Himalayas. A student of the Graduate School of Engineering at the University of Tokyo, he accepted a job offer from Chugai in 2001, 'because they gave enough holidays for cycling.' The norm in Japan is just one week off work at a time; Chugai allowed three.

He knew nothing at all about antibodies.

In retrospect, Igawa muses, his total ignorance of antibody biology was an asset, because it freed him to think of Hattori's candidate molecule as just that – a molecule that could be tinkered with and adjusted. The rest of the team were mostly biologists, all somewhat in awe of the complexity of antibodies. They were reluctant to mess around with these intricate structures; Igawa had no such qualms.

Joining the haemophilia programme in 2004, Igawa was paired with the team's other engineer, Hiroyuki Tsunoda, to optimise and humanise the lead candidate emerging from the second generation of bispecific antibodies. *Optimising* means refining the molecule to make it more potent and improve its pharmacokinetics and bioavailability. *Humanising* is a process of tweaking the molecule, which comes from a mouse, to be more like a human antibody and therefore less prone to attack from the human immune system.

The two engineers were also tasked with solving the chain association problem. If they couldn't do that, it didn't matter how good the molecule was: Chugai would not be able to manufacture it efficiently enough for a commercially viable medicine.

Igawa and Tsunoda studied the problem. Hattori's bispecific antibody consisted of four polypeptides – two light chains and two heavy chains – which were synthesised and then bound together inside a cell. The cell assembled the chains randomly – any two light chains with any two heavy chains – so that only one of the ten possible assemblages was actually the candidate molecule. The other nine combinations were useless; those molecules would have to be separated out and removed before the tiny quantity of clotting antibody could be used. As the ten different combinations were all very similar, separation and purification would be nearly impossible.

Far better if they could have the cell produce *only* the clotting antibody.

Tetsuo Kojima had already seen an idea put forward by Paul Carter at Genentech that he reckoned could solve part of the problem, and he had shared it with Hattori at the beginning of the bispecific antibody project.[17] Each of the two binding sites on a bispecific antibody is made up of a bit of a light chain and a bit of a heavy chain. But in certain cases, most of the binding is actually done by the heavy chain; the contribution of the light chain is low. So, in theory, one might be able to use the *same* light chain on both binding sites without spoiling the binding specificity for Factor IXa on one and for Factor X on the other.

The two binding sites on the bispecific antibody had been generated separately by lymphocytes in two different mice, so it wasn't surprising the light chains in each were different. But that didn't mean they had to be.

Kojima, Miyazaki and Soeda successfully identified a common light chain that would work on both sides of the mouse-derived bispecific antibody. It took Igawa and Tsunoda a further six months to humanise the common light chain without diminishing the *in vitro* activity of the bispecific antibody. By re-engineering the light chain genes, they were able to create recombinant cells that synthesised the original two heavy chains and a single, new light chain. Now, there were only three possible combinations:

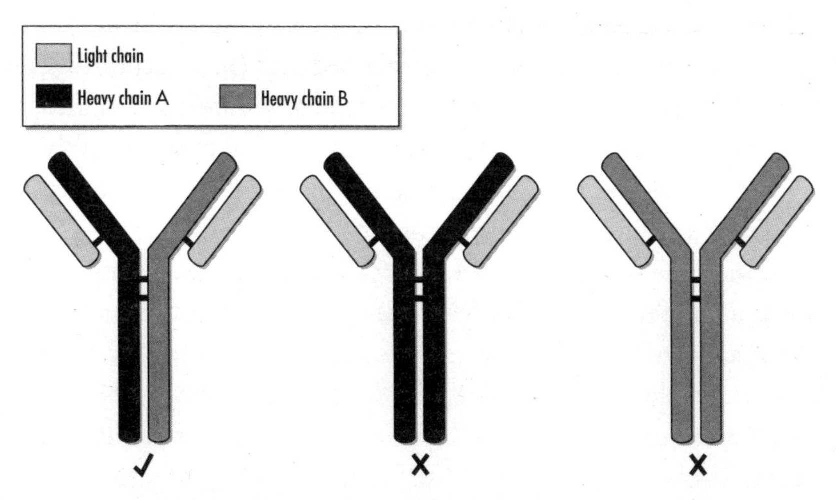

Figure 14. The several ways that two types of heavy chain and one common light chain can assemble.

There were two ways to assemble the first – correct – combination from the three polypeptide options, so two out of four molecules formed, or fifty per cent of the yield, would now be the clotting antibody.

It was a huge step forward, and Igawa and Tsunoda already had an idea for how to push the yield up even further. Hattori's dream was becoming a viable possibility at last.

'We were very confident,' says Hiroyuki Saito of the mood in 2006. The candidate molecule, Ab-x, had high potency, binding tightly to both Factor IXa and Factor X. *In vivo* tests seemed to go well. The team had established a method to induce haemorrhagic symptoms in haemophilic animals. Two animals treated with Ab-x had less serious haemorrhagic symptoms than two control animals.

At last, Chugai's executives were convinced. They declared the haemophilia A programme their priority research project and greenlighted clinical trials. Expectations were sky-high.

Saito was appointed project leader and put in charge of taking Ab-x from lab to clinic. He moved to Chugai's headquarters in Tokyo, relocating with his wife from the quiet environs of Mount Fuji to the clamour and bustle of the nation's capital. With the project team, he designed the trial protocol and established relationships with the physicians who would oversee the first phase of trials.

Meanwhile, the molecule was in the hands of Chugai's toxicology department for its pre-clinical tests.

It took a few months for the chronic tox study to be finalised, and when the report eventually landed on Saito's desk it was devastating. Elevated doses of Ab-x caused heavy bleeding in animal muscles, bladder, testes and heart. Far from improving coagulation, Hattori's clever new concept had delivered a drug that actually made bleeding worse – at least at high dosage in non-haemophilic animals.

At first, no one could understand it. Why would an antibody that mimicked the function of Factor VIII *prevent* coagulation? It made no sense. Saito was floored. He had worked on the programme for five years. He had moved house for it. And now he had no choice but to terminate it. 'It was very painful,' he recalls.

Although Chugai understood the value of learning from failure, and recognised that failure was built into the drug discovery process, hard questions were nevertheless asked of Saito and his colleagues by senior executives. Why hadn't they spotted this problem earlier? Why couldn't they explain how their wondrous haemophilia drug actually made bleeding worse?

Saito's boss was more understanding. 'He asked me if I wanted to stay.' But by that point, there was little left for him to do. They agreed that Saito should be transferred to a different team, in Neurology.

He was off the haemophilia programme.

* * *

Kunihiro Hattori had always longed to invent something remarkable. 'I strongly wanted to create an epoch-defining new drug,' he remembers. Now he was forty-six, facing another failed project, and he feared he'd never make it. To start from scratch on a new concept might require fifteen years of R&D. 'The disappointment was very bad.'

For a few days, it looked like the bispecific antibody project was finished. Hattori had no choice but to reassign his two engineers, Igawa and Tsunoda, to other projects. But the remaining team members were eager to keep going, and Hattori decided to try again. He was determined to make a better molecule.

Senior management was not supportive of this new attempt. Grudgingly, they agreed to give the Coagulation Research Group one last chance. Hattori's team would have just eighteen months – until midnight on 30 June 2008 – to find a candidate molecule. Not a second longer. After that, the programme would be shut down for good.

'I had never had such an exact deadline before,' laughs Kitazawa.

A year and a half sounds like a long time, but it's nothing in drug discovery. Hattori, Kitazawa and their colleagues were going to have to work fast and work hard.

But first, they had to figure out why Ab-x exacerbated bleeding at higher doses.

They knew that Ab-x was a good antibody: it bound tightly to both targets, Factor IXa and Factor X. Closer investigation revealed this to be the root of the problem. In binding too strongly to Factor IXa, Ab-x effectively blocked Factor IXa from binding to Factor VIIIa at higher doses. Thus, Ab-x was disrupting its target's function rather than simply linking it to Factor X. The high affinity that the team had assumed was a beneficial attribute of the molecule turned out instead to be its downfall.

A less good antibody would make for a better medicine, they realised. It was a counter-intuitive but vital insight.

By now, Takehisa Kitazawa had grown into a leadership role in the thrombosis and haemostasis group, and he accelerated the search for a viable bispecific antibody. This time, the team prepared around two hundred antibodies to Factor IXa from mice, rats and rabbits, and another two hundred to Factor X, ultimately yielding some forty thousand bispecific antibodies with a wider genetic diversity. After months of screening, just ahead of their deadline, the team selected their best molecule to take forward for further refinement. It bound to Factor IXa and Factor X, but had much less coagulation-blocking activity than Ab-x.

When I first read the paper describing this effort, I was completely blown away.[18] Screening thousands of molecules is routine when developing a small molecule, but it's a very different matter to screen complex biologics. Antibodies are proteins, which can easily be denatured and damaged if handled incorrectly. Each one has to be grown in cells, purified and stored at the right temperature and in the right medium. Compared to small molecules, which can be ordered from chemical manufacturers and stored for long periods in ambient temperatures, biologics are a logistical headache. I had never heard of anyone screening tens of thousands of antibodies before!

Tom Igawa returned to the team in time to optimise and humanise the molecule. With Hiroyuki Tsunoda, he had by now established a technique for solving the chain association problem in its entirety. The candidate molecule consisted of two identical light chains and two different heavy chains. These three types of chains were manufactured in a recombinant cell, and then assembled randomly. In that random assembly, some antibodies would be built with two of the same type of heavy chain (AA or BB), producing useless molecules that would have to be separated out from the final medicine.

The challenge was how to stop two identical heavy chains coming together. The solution, Igawa and Tsunoda concluded, lay in an electrical charge.

Some molecules carry a tiny electrical charge, either positive or negative. Like magnets, two molecules of the same charge will repel each other, while molecules of different charges will attract each other. Metaphorically speaking, Igawa and Tsunoda decided to strap a magnet's 'north pole' to one heavy chain and a 'south pole' to the other, ensuring only different heavy chains would come together.

They engineered a very small alteration in heavy chain A to give it a negative charge at the point where it would bind to the other heavy chain. They gave a positive charge to heavy chain B. Opposites attract, so heavy chains A (negative) and B (positive) were drawn to each other as the antibodies assembled. Meanwhile, identical heavy chains, having the same charge, repelled each other.

With this clever innovation, they ensured ninety-nine per cent of the antibodies assembled contained one of each heavy chain.

Moreover, because they had given each heavy chain a different charge, the *effective* antibodies had a different overall charge (approximately neutral) to the *by-product* antibodies (very positive or very negative). We can separate molecules with different electrical charges using a process called ion exchange chromatography. In this way, the by-product impurities could be easily eliminated.

By tweaking the electrical charge of the heavy chains, Igawa and Tsunoda had almost doubled the manufacturing yield *and* made purification easier.

Thanks to the groundbreaking work of Tom Igawa, Hiroyuki Tsunoda, Paul Carter and others in the field, the chain association problem is now largely solved, opening the door to many more possible bispecific antibody drugs.

After eighteen months of hard work to improve potency and pharmacokinetics, during which another 2,400 bispecific antibodies were engineered and tested, the final molecule was ready.[19] It was labelled **ACE910** (***A****ntibody mimicking* ***C****oagulation factor* ***E****ight, connecting coagulation factors* ***9****a and* ***10***). This time, the preclinical toxicology tests came back clear. Hiroyuki Saito was brought back from Neurology to resume his project leadership, taking ACE910 into clinical trials. 'That was the highlight of my career,' he says happily.

Midori Shima, by then Professor of Paediatrics at Nara Medical University, helped set up and run the Phase I studies. For the first study, sixty-four healthy male volunteers were given a single dose of the drug subcutaneously at Showa University in Tokyo. ACE910 was found to have a half-life in the bloodstream of four to five weeks, meaning patients would need far less frequent injections.[20] 'I was very surprised,' says Shima. 'I was expecting maybe a week.' No serious adverse effects were detected. Most importantly, Factor VIII function-mimetic procoagulant activity could be seen in the volunteers' blood.

Now in his sixties, Shima still manifests a youthful delight as he thinks back to that first human result. He's a man who enjoys life, particularly cooking and the craft of smoking meat that he picked up while living in San Diego in the 1980s. He's been researching and thinking about Factor VIII for decades. Witnessing an antibody mimic the action of Factor VIII in humans was an experience he could never have imagined.

In general, Phase I studies (other than in cancer) focus only on the safety and pharmacokinetic properties of the experimental drug in *healthy* individuals. However, for ACE910, Shima and his colleague Associate Professor of Paediatrics Keiji Nogami* agreed with the Chugai team that they would perform a Phase I study in haemophilia patients, too. Eighteen Japanese men with severe haemophilia A were given weekly subcutaneous doses, and the efficacy of the drug in preventing bleeding was evaluated.

The first subject was one of Shima's patients from Nara Medical. Although Shima and Nogami were confident that ACE910 would have no serious adverse effects, they weren't at all sure it would be efficacious. Shima admitted to being very nervous about the outcome, as was the patient and the whole Nara Medical team. To take part in the study, patients were required to give up their

* Also of Nara Medical University.

existing clotting management regimen. That meant they had to stop using Factor VIII replacement therapy, potentially leaving themselves at risk of serious bleeds.

The team's concern soon turned to confidence, however, because the patient's intermittent bleeding ceased from the first dose of ACE910 onwards. No serious safety concerns arose. Similar efficacy and safety were seen in the other study subjects, including those with inhibitors (the immune response against replacement Factor VIII).[21]

I learned in depth about ACE910 around this time when my group at Roche pRED was preparing to help transition it from early phase development at Chugai to late phase development at Roche and Genentech.* Reviewing the way Chugai had used an antibody to replace Factor VIII, I thought, 'How is that even *possible*?'

And yet the results showed it was possible, as exemplified by Shima's long-standing patient, Kenji, who also participated in the study. 'Kenji was very nervous for the first three months,' recalls Shima. But the evidence was clear. With only one leg and a fear of bruising, he would usually attend his hospital appointments in a wheelchair. Now, he started walking everywhere again. After three months of reduced pain and internal bleeding, he felt confident to take a hot bath. And he began going to the gym for the first time.

'He started muscle training,' laughs Shima. 'He's a very big man and his muscle increased a lot! He became quite a different person, like a macho man, from the one who came to the hospital in a wheelchair.'†

Shima and Nogami were delighted by the results. The *New England Journal of Medicine* accepted their article for publication, a real triumph for the physicians and validation of the clinical importance of their findings. Such international recognition is never guaranteed, and the *NEJM* is among the most respected

* Members with Chugai of the Roche Group.

† Chugai would like to state that this is one specific case and the same effect is not guaranteed for all patients.

peer-reviewed medicine journals in the world. In Japan, the government and media celebrated the study as a showcase example of a joint industry-academia project.

It was time to name the drug. The INN suffix for a humanised monoclonal antibody applying to circulatory diseases is *-cizumab*. Zenjiro Sampei, who had worked with Igawa on the molecular engineering for ACE910, proposed *emi*, a Japanese word for *smile* that also abbreviates [Factor] ***e**ight **mi**metic*. Thus, *emicizumab* was born.

The pivotal Phase III trial was conducted internationally, in partnership with Roche and Genentech. Gallia Levy led the clinical trial team. 'It was a huge deal for the haemophilia A community,' she recalls, particularly for people with inhibitors, who had so few pharmaceutical options. A physician and molecular biologist, she felt under real pressure to get it right for them. 'The data from Chugai for Phase I looked good,' she says. Now it was up to her team to demonstrate to regulators around the world that emicizumab was an effective drug with no serious adverse side effects.

The biggest challenge, she soon realised, was the variable nature of patient reporting. Trial investigators would be administering emicizumab to people in multiple countries as a prophylactic, and then relying on them to report when bleeds occurred. Internal bleeding is not always obvious. Some people feel pain, or tingling, or warmth, when they bleed internally; others would say, 'I know I'm bleeding because I get an *aura*.' People with haemophilia are not always sure whether they are bleeding internally. Different study participants will report bleeding events differently, according to what they have been taught, or the diagnostic techniques they have developed for themselves. Such variability in reporting makes it very difficult to run a proper controlled study, comparing a placebo arm to a treatment arm.

So Levy started with a non-interventional study, asking participants to record their bleeding history over a period of time *without* emicizumab. Each participant thus built up a personal profile of their normal bleeding pattern before being invited to join the emicizumab Phase III study. In addition to the standard control arm, each subject's prior profile – recorded with their own biases – could serve as their own individual control. If they reported less frequent bleeding once dosed with emicizumab, that would be evidence of the drug's efficacy. This approach had never been used before in a registrational haemophilia study.

The new approach worked so well that the FDA has now changed its guidance, recommending such 'intra-patient comparison' studies for other haemophilia trials. 'People tend to think the gold standard is a randomised trial,' says Levy. 'Well, in this case, a randomised comparison wasn't really the best option. So we have sort of changed the gold standard for haemophilia trials.'

Levy was particularly eager to make emicizumab available for children as quickly as possible. Living with haemophilia is even tougher for kids, she pointed out. Regulators require separate studies in children under twelve. Levy argued that the *efficacy* seen in adults could be extrapolated to children, and the FDA accepted that argument; the team would only have to show the drug was *safe* in children, without having to prove efficacy to the usual exacting standards. They were able to do this with an accelerated study in twenty children, and won paediatric approval at the same time as adult approval, an unusual achievement in pharmaceutical development.

Emicizumab was approved for use by people with inhibitors in the United States at the end of 2017, under breakthrough therapy designation granted by the FDA.[22]

For Kunihiro Hattori, the successful development of emicizumab represented his crowning achievement in life. 'I have made my contribution to human society. I'm ready to die any time – I have nothing left to do.'

Takehisa Kitazawa doesn't think Hattori's ready to retire: 'Personally, I expect Hattori-san to do more.' And for Kitazawa, who worked on the haemophilia project for twenty years – longer than anyone else – their joint achievement has been life-changing. 'I've heard from patients, and doctors, and patients' families, especially mothers . . . It has been extremely moving. I'm so proud of Hattori-san and all my colleagues.'

At fifty-three, Kitazawa, too, might be expected to rest on his haemophilia laurels, but he has other ideas. He isn't done yet. 'My present dream is to create an epoch-making drug,' he declares.

CHAPTER 5

Therapy-Resistant Breast Cancer

Condition: Breast Cancer
Target: PI3K
Gene: *PIK3CA*
Innovation: Alpelisib [al-PEH-li-sib]
Company: Novartis

LEW CANTLEY WAS NOT ACTUALLY INTERESTED in cancer when he began studying cell membranes. A chemist who ended up working as a biologist at Harvard at the tail end of the 1970s, Cantley would insist that everything in biology could be explained by chemistry. In his view, biologists often couldn't see the truth in front of them because they weren't looking small enough: they would talk about chemical signals passing through the cell membrane, but they wouldn't ask themselves how exactly that happened. To truly understand a cell's biology, Cantley argued, you had to find out what was happening at the molecular level.

Fascinating things happen around the cell membrane. Our cells are contained and defined by this complex layer. It is permeable to some substances, impermeable to others. Its permeability may vary depending on what chemicals are circulating in the blood. Messages from one part of the body to another are transmitted across the membrane. Everything our cells need has to pass through it. That includes our most basic need of all – energy, in the form of glucose.

It was glucose, and its selective uptake into cells, that held Cantley's attention. He knew that insulin regulated the amount of glucose absorbed into a cell, but how exactly did that happen? What was the molecular mechanism underlying one of the most vital processes in all of life?

Cantley is a chemist to his bones. As a young boy, he learned how to make gunpowder from three simple ingredients he could buy from the local store. Fireworks were illegal in his home state of West Virginia, so he built his own. 'One of my presents growing up was a chemistry set, and I loved mixing things together and seeing colours change and causing occasional explosions. I just found that amazingly fun.' But the most interesting application of chemistry, for the son of a farmer, was life itself. He longed to know how the reactions he saw taking place in test tubes translated into biological processes.

Specifically, Cantley wanted to establish how insulin regulates the passage of glucose through the cell membrane. Insulin activates the sodium–potassium pump, a molecular structure which moves ions (electrically charged particles) across the membrane, and he found that this pump depended for its function on a set of surrounding lipids (fats) known as ***p**hosphatidyl**i**nositol* or *PI*. Cantley discovered that the behaviour of the sodium–potassium pump could be altered by phosphorylating (adding a phosphate group to) the surrounding PI. He speculated that insulin might be stimulating the phosphorylation of PI and thus regulating the activity of the sodium–potassium pump. If so, that meant insulin worked by activating some kind of *PI kinase*.

You remember kinases? EGFR is a kinase. To recap, a kinase is an enzyme that adds a phosphate group to a molecule. This process, known as *phosphorylation*, changes the properties and behaviour of the molecule.

There was so much to explain about EGFR and its mutations in Chapter 2 that we glossed over how exactly it causes cell proliferation. As we take another look at cancer, it's time to go deeper into the biology of kinases.

Kinases phosphorylate. That's what they do. Phosphorylation performs a function a bit like turning a switch on or off, and it's incredibly widespread. In every cell in our bodies, innumerable molecules are being phosphorylated and dephosphorylated (where phosphate groups are removed) all the time.

One phosphorylated molecule often phosphorylates hundreds more, and each of them phosphorylates hundreds more, and so on. In this way, a cascade of signals is created. At each step in the cascade, the signal is amplified, so that a process triggered by a single phosphorylation event results in thousands of activated molecules further down the chain.

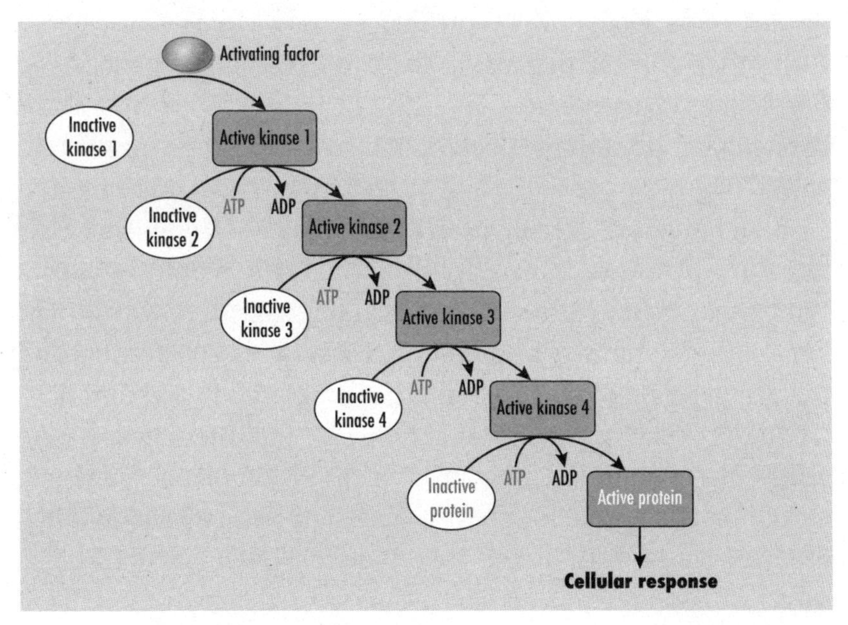

Figure 15. A generic cellular signalling pathway.

We call this a *signalling pathway*. It is the mechanism by which many of the processes in our bodies are controlled. In summary, *kinases* are enzymes that catalyse *phosphorylation*, and phosphorylation is a common mechanism cells use to switch cellular and molecular processes on or off.

One such process is cell division.

What exactly is cancer? What unifies diseases as dissimilar as breast cancer and leukaemia, which affect the body in radically different ways? A simple definition of cancer is excessive cell proliferation: abnormal cells divide rapidly and may invade nearby tissue or spread to other parts of the body. This is as true of blood cancers like leukaemia as solid tumour cancers like breast cancer. In every case, cells divide more than they should.

Most of the roughly 30–40 trillion cells in our bodies live for only a few weeks or months, meaning that they have to be replaced regularly. Consequently, billions of cells are killed off every day, and billions more are created through a process of cell division – a mature cell divides into two identical daughter cells. As you can imagine, it is vitally important that these two processes balance each other. Kill off more cells than you create, and tissues and organs start to degenerate (we see this in ageing). Create more cells than you kill off, and the likely result is cancer. If those extra cells spread to other parts of the body, you have metastatic cancer.

Both processes – cell division and cell death – are controlled by signalling pathways. Kinases play pivotal roles as on-off switches throughout these pathways. If something goes wrong with the kinases, the result may be an imbalance between cell division and cell death, leading to uncontrolled cell proliferation and cancer.

This is what happens when the *EGFR* gene is mutated in lung cancer: the tyrosine kinase switch it encodes becomes permanently locked 'on'. The signalling pathway instructs cells to keep dividing, producing new cells faster than they are eliminated. Like the sorcerer's apprentice faced with his unstoppable magic broom, the body cannot halt the process of abnormal cell proliferation.

Many different signalling pathways can cause – or fail to suppress – cell proliferation. But what we learned from the case of mutant EGFR is that it might be possible to treat certain cancers if we can disrupt the right signalling pathways.

Fans of *The Lord of the Rings* may remember the warning beacons of Gondor, a chain of great fires on mountaintops that could be lit to summon aid in time of war. That was a signalling pathway: the first fire is lit; watchmen on the next mountaintop see it and light their own, prompting the next set of watchmen to light their fire, and so on. The message is carried from mountaintop to mountaintop in sequence. If an enemy wished to prevent the message getting through, they might scale any one of the mountains in advance of the fire-lighting and incapacitate the watchmen on that peak. The rest of the signalling pathway remains in place, but if that fire is not lit the message will not get through.

In the same way, researchers seeking a treatment for cancer can attempt to knock out one of the molecules in the relevant cell proliferation signalling pathway. The oncogene (cancer-causing gene) encodes a mutant protein which activates a signalling pathway leading to cell division . . . but if a crucial pathway molecule is incapacitated by a drug, the message never gets through. That pathway molecule is often a kinase.

Although Lew Cantley was focused on insulin when he dreamed up the idea of a PI kinase, he was well aware of an important link between his work and cancer biology. One defining feature of a tumour cell is that it consumes glucose at a much faster rate than normal cells: for tumours to grow, they require a great deal of energy. There is a clear parallel between insulin-related diseases like diabetes, where too little glucose is absorbed by cells, and cancers, where tumour cells need excessive glucose. Perhaps, he speculated, the same control mechanism was going awry in both cases.

Cantley had followed the exciting developments in cancer research through the 1970s, when the first oncogene, *SRC*, was discovered and shown to encode a kinase. He paid particular attention to a subsequent discovery in 1983, when Ray Erikson reported a new type of *lipid* kinase (a kinase that phosphorylates lipids rather than proteins) associated with the SRC kinase. Cantley hypothesised that it might be the PI kinase he was looking for.

To test his hunch, he set up an experiment, adding the cancer-causing SRC kinase to PI and showing that this led – presumably through the action of another kinase – to the phosphorylation of PI. Later, his lab was able to purify this new PI kinase and confirm that it is the 'switch' that insulin activates (via the insulin receptor) to control glucose uptake. Further investigation showed that, of the many genes carried by an oncogenic virus, only those encoding proteins that could activate Cantley's PI kinase actually turned cells cancerous. The hypothesised link between this newly discovered enzyme and cancer was real.

'I went home and opened a bottle of champagne and told my wife we'd made a huge breakthrough,' recalls Cantley.

Different kinases add phosphate groups to different parts of the PI molecule, and some of these were already documented. But Cantley's PI kinase did something never seen before: it added a phosphate group to the 3-hydroxyl position of PI's inositol ring. Chemists who had worked on PI for twenty years refused to believe this *phosphatidylinositol 3-kinase* or *PI3 kinase* actually occurred in nature. 'They said, "It doesn't exist!"' Cantley laughs ruefully. For two years, nobody outside his laboratory believed him. It would be four years before he could obtain any grants to work on the enzyme.

Yet within a decade, PI3 kinase, or *PI3K*, would become one of the hottest targets in all of cancer biology.

Lew Cantley understood the importance of PI3K to cancer medicine: he recognised that a drug able to disrupt PI3K's activity – a *PI3K inhibitor* – might well slow or stop a range of different cancers. But he did not believe PI3K was a practical target for a cancer drug. It was simply too important to insulin control of glucose uptake. 'I didn't think you could thread the needle,' he admits. 'I couldn't see how you could hit PI3K without starving normal cells of glucose.'

Nevertheless, he kept working on PI3K with colleagues and collaborators, painstakingly mapping out the many signalling pathways that lead to the enzyme. PI3K can be activated by dozens of other kinases. *SRC* is just one of the genes that ultimately triggers PI3K. So PI3K sits at a junction linking multiple signalling pathways that control and regulate a variety of aspects of cell growth, metabolism and death. It took Cantley and other researchers more than a decade to figure out the details.

By that time, pharma companies were zeroing in on Cantley's kinase. As the world prepared to welcome the new millennium, the concept of a PI3K inhibitor was one great pharmaceutical hope on which a thousand dreams of curing cancer were pinned.

Novartis had a particularly compelling reason to investigate PI3K. The Swiss pharma giant was already running clinical trials of a molecule that would soon come to be seen as a cancer wonder drug. Imatinib was approved by the FDA in 2001 for treatment of a blood cancer called chronic myelogenous leukaemia (CML). Almost every patient treated in the trials showed remarkable improvement, their white blood cell counts swiftly returning to normal levels. Survival rates were close to ninety per cent, earning imatinib sobriquets like 'magic bullet' and 'miracle drug'.

I was in training as a medical oncology fellow when imatinib was first approved. Back then, if chemotherapy or bone marrow transplants did not work, patients with refractory CML had no options. Suddenly this pill came along and dramatically increased survival rates – we were stunned. Cures for all kinds of cancers seemed just over the horizon.

Imatinib's story has been told elsewhere,[1] but what matters for us is how the drug works.

It is a kinase inhibitor.* So Novartis already had solid proof that blocking the activity of a kinase within a signalling pathway was an effective, targeted way to treat cancer.

Early attempts to identify kinase inhibitors through traditional screening methods were largely unsuccessful. But in 1977, a natural poison called staurosporine was discovered in a soil sample

* Imatinib inhibits the ABL kinase in patients with CML who harbour BCR-ABL kinase fusions.

analysed by investigators in a Japanese medical research facility. Synthesised by bacteria, staurosporine turned out to be a potent kinase inhibitor, blocking multiple kinases in critical pathways, leading to cell death.

Staurosporine is not safe for humans, but researchers used it as a chemical model to identify more selective kinase inhibitors. Few organisations at that time did more work in this area than Ciba-Geigy, one of the Swiss pharma companies that would eventually merge to form Novartis. Through the 1980s and 1990s, a team led by Alex Matter and Nick Lydon discovered dozens of chemical compounds that inhibited one or more of the roughly five hundred kinases in the human body. Gradually, they built up a library of kinase inhibitors that might one day prove useful against diseases as yet unspecified.

By 2000, Novartis had a plausible new cancer target in the PI3K signalling pathway. It had a large collection of kinase inhibitors. And it had a track record of treating cancer with kinase inhibitors. Now it needed the right team to put it all together.

Michel Maira has researched cancer all his professional life. It has, he says, a 'fascinating biology'. He became interested in the role of signalling pathways in tumour formation long before he joined Novartis. As a young molecular and cellular biology PhD candidate, he spent over four years studying how one well-known oncogene transforms normal cells into cancer cells. For his postdoctoral training, he worked in the laboratory of a Swiss biomedical researcher whose principal interest was the human *kinome* – the collective term for all the genes in the human genome encoding our kinases. He was tasked with investigating a key component of the PI3K pathway called AKT.

Found downstream of PI3K in the signalling pathway, AKT is another important kinase; when activated, it promotes cell proliferation and hinders cell death. If PI3K becomes permanently switched on, it causes permanent activation of AKT. Uncontrolled cell growth is the inevitable result.

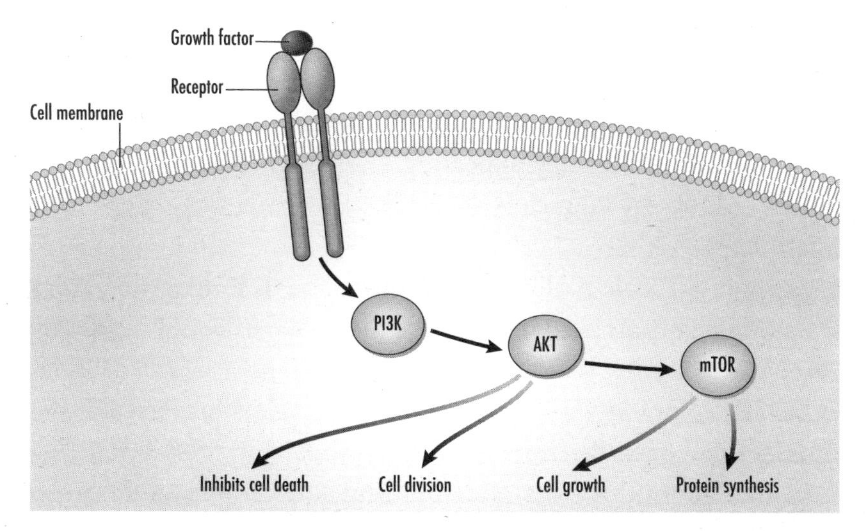

Figure 16. The PI3K-AKT signalling pathway.

Fascinating as the work was, Maira was not content to stay in academia. 'I really wanted to go to a pharma company because that way I knew I could have a very strong impact on cancer biology understanding, and establish potential new therapeutic avenues.' Like many of us, Maira had seen members of his own family struck down by cancer.

Meanwhile, Novartis was looking for a way to disrupt the PI3K pathway, and had concluded that AKT was a more promising target than PI3K itself.* Think of that chain of mountaintop fires – if you want to break the chain, you have a choice of fires to take out; you'll probably choose to scale the least vertiginous mountain, or the one with the sleepiest watchmen. Functionally, it may not matter where you break a chain of communication, so long as the signal can't get through.† Novartis chose to target AKT, and went looking for scientists who understood it. They found Michel Maira in 2001.

* AKT is a protein kinase, while PI3K is a lipid kinase. At the time, protein kinases were much better understood, and therefore considered easier to work on.

† This is a simplification for the sake of clarity. Cancer biology is remarkably complicated, and multiple oncogenes and signalling pathways are often involved in any one cancer. This means that an activated oncogene may not on its own lead to cancer, but it also means that more than one pathway or oncogene may need to be targeted simultaneously to treat a particular cancer.

Maira was delighted to join Novartis and gain access to one of the world's largest collections of kinase inhibitors. Nothing like it was available in academia. Moreover, the cultural environment at Novartis was immensely appealing. 'This was the golden age for targeting kinases,' he recalls wistfully. 'There was huge hype after imatinib.'

Appointed lead biologist on Novartis's AKT drug discovery team, Maira began preparing assays to test the effect of candidate compounds on AKT in living cells. Lead chemist Carlos Garcia-Echeverria worked with molecular modelling expert Pascal Furet to select compounds from Novartis's proprietary collection that seemed a good fit for the AKT binding site. Furet also designed some analogue compounds, which Garcia-Echeverria synthesised. When these various compounds were added to model biochemical systems in a test tube, AKT stopped functioning, exactly as hoped.

But then it was Maira's turn to test the candidate compounds in his cell assays – and they didn't work. When administered to a living cell, each of the chosen molecules proved to be a dud. Some broke down too swiftly, others just never seemed to reach their target. After two years of hard work, the AKT team had to admit defeat.

Unwilling to give up on the promise of the PI3K pathway, Maira and Garcia-Echeverria pivoted to a different part of it. As one mountaintop proved inaccessible, they began to scale another. Furet had suggested that a particular series of compounds identified in the AKT project might in fact be PI3K inhibitors, based on their apparent fit with the PI3K binding site. And so, PI3K itself became the target.

To complicate matters, PI3K occurs in several different structural forms, or *isoforms* (alpha (α), beta (β), gamma (γ) and delta (δ)), with each isoform encoded by different genes. Maira and Garcia-Echeverria set out to find a *pan-PI3K inhibitor* – a chemical that would block all forms of PI3K.

This decision was sensible at the time, but in hindsight, it would set back the search for a safe and efficacious PI3K inhibitor by many years.

The first years of the search for a pan-PI3K inhibitor were marked by optimism. Maira, Garcia-Echeverria and Furet identified a series of chemical compounds that inhibited PI3K in all its forms, and also unexpectedly inhibited another critical kinase further downstream in the cell creation pathway called *mTOR* (see figure). The over-activation of mTOR has been strongly linked to tumour growth in a number of different cancers.* The new candidate drugs could, as it were, take out not one but two of those mountaintop beacon fires, an extra-certain defence against cancer.

The most promising of the dual PI3K/mTOR inhibitors were BEZ235 and BGT226. Not only did they work in biochemical assays, but both molecules showed effective inhibition of PI3K, AKT and mTOR in Maira's living cell assays. The whole PI3K-AKT-mTOR signalling pathway that drove the growth of numerous different tumours could simply be switched off.

It felt like they'd found the next imatinib.

Around the same time, Novartis made a significant acquisition, buying a Californian biotech company called Chiron. Best known for its vaccines and diagnostics divisions, Chiron also had a biopharmaceutical arm, and one of its research interests was PI3K. They too had identified a pan-PI3K inhibitor, BKM120, and this asset now passed to Maira's team.

Unlike BEZ235 and BGT226, BKM120 did not impact mTOR, and Maira could see that this difference might have advantages. mTOR performs a critical role in a lot of cellular processes, so a dual inhibitor could bring unwanted side effects by inadvertently switching off processes essential to normal cell function. Adding a single-action PI3K inhibitor to the toolbox insured against that concern.

* Novartis would successfully launch an mTOR inhibitor, everolimus, as a solid tumour cancer treatment a few years later.

So now Maira was investigating three molecules, all showing great results in cell assays and animal models. All three were able to shrink tumours in mice. These chemical compounds unquestionably worked against cancer. But until now, none had been tested in humans.

Candidate molecule	Name	Company of origin	Enzyme targeted	PI3K isoforms targeted
BEZ235	dactolisib	Novartis	PI3K, mTOR	All
BGT226	-	Novartis	PI3K, mTOR	All
BKM120	buparlisib	Chiron	PI3K	All

Table 1. List of PI3K inhibitors in development at Novartis.

First-in-human studies are always an anxious moment in drug development. Prior to this point, a lot of pre-clinical work – in petri dishes and in animals – is performed, but no laboratory model can completely mimic the human body. As we've seen, we can never predict exactly what will happen when we put molecules into humans for the first time. More than ninety per cent of candidate molecules fail at this point. Some turn out to be unacceptably toxic, others just don't work well enough against the disease. For BEZ235 the problem was a little more complicated.

The dose of any medicine is vitally important. It must be high enough to be effective but low enough to avoid problematic side effects. Pharma companies devote a great deal of time and resource to establishing the right dose for any candidate molecule.

For BEZ235, that proved impossible because the pharmacological activity of the compound varied so much from person to person. In one patient, the molecule seemed to reach every tumour cell and inhibit mTOR and PI3K effectively, while in another it would barely show up. Its pharmacokinetics – the way the compound behaves in the body – were extraordinarily variable. In fact, the variability even occurred within the same individual – a dose one day would behave quite differently to a dose in the same patient two weeks later. That made BEZ235 unpredictable, unreliable and possibly dangerous. Unless the variability could be fixed, no physician would ever know what dose to give their patient.

Months went by as the team tried to understand what was causing the pharmacokinetic variability. At first, they believed it was down to the poor solubility of the compound. They tried formulating it in different ways, even switching from pills to granules sprinkled on cereal. Nothing worked. As it turned out, solubility wasn't the issue.

Many drugs are metabolised primarily by a single mechanism, often an enzyme in the liver. This makes it relatively straightforward to predict how quickly they will be broken down. But BEZ235 is metabolised by enzymes (aldehyde oxidases) scattered throughout the body, and this complexity was found to drive the variability. Years of pre-clinical studies had demonstrated the promise of BEZ235 as a potent anti-cancer drug, but that pharmacokinetic unreliability in humans undermined it all.

Novartis pulled the plug on BEZ235.

BGT226 was a close molecular cousin of BEZ235 – in medicinal chemistry parlance, it had the same *scaffold*. So it was not surprising to find it also had pharmacokinetic problems. But the bigger problem revealed by the first human trials was toxicity. Where BEZ235 had been well tolerated by patients, BGT226 made them nauseated and sick.

Patients with advanced cancer are willing to put up with much more unpleasant side effects from their medicines than most people, so it tells you a lot about the toxicity of BGT226 that clinical trials were swiftly halted. Was the toxicity due to the dual action of the molecule? Was it asking for trouble to interfere with mTOR as well as PI3K? The team never did find out what the root cause of the toxicity was.

BGT226, too, was history.

It was a rough few years for Michel Maira, Pascal Furet and Carlos Garcia-Echeverria. Both the dual-action inhibitors they'd discovered had failed. But they still had Chiron's molecule, BKM120. Would a single-action PI3K inhibitor fare any better?

At first, everything seemed to go well. Initial clinical studies showed the molecule was dispersing effectively through patients' bodies and was being metabolised at a steady rate – its pharmacokinetics were good. The Phase I trial concluded with good evidence of efficacy and no overly troubling side effects.

Now named buparlisib, Chiron's molecule had become a really important bet for Novartis. The team were increasingly confident they had a winning cancer treatment on their hands, and they wanted to bring it to market as soon as possible, ideally with FDA approval for the treatment of a wide range of cancers. That meant recruiting patients with all the different cancers of interest for the next trial. It became a very large and complex programme.

But something troubling happened in that trial. Patients began reporting depression, sometimes severe, and other mood disorders. Their doctors and families noticed worrying changes in their character.

These neuropsychiatric side effects – none of which were detected in the preclinical safety studies – pointed to a direct impact of buparlisib on the brain. When the Novartis team investigated, they discovered the drug was able to cross the blood-brain barrier. As we saw with osimertinib, this can be a valuable attribute if we want to treat brain metastases or other disorders of the central nervous system. But if the brain is not the target, it's much safer to use drugs that can't cross the blood-brain barrier.

Trial investigators also reported multiple cases of hyperglycaemia – blood glucose levels were too high. It was precisely the issue that Lew Cantley had predicted: PI3K is critical to the uptake of glucose into the cell, so a drug that disrupts it may cause a major glucose imbalance.

Theoretically, the hyperglycaemia challenge could be managed, but the neuropsychiatric effects were too great a risk. And so, in 2016, Novartis took the painful decision to close the entire pan-PI3K inhibitor programme.

Michel Maira had worked on PI3K inhibitors for more than a decade. He would continue to research cancer therapeutics at Novartis, and has gone on to make exciting breakthroughs in controlling another oncogene, but his massive personal investment of time, skill and passion in PI3K ended in unimaginable disappointment.

'It's never easy to face the situation where you see things are not going the way you want,' he says. 'But what I've learned in more than twenty years of drug discovery is that this is part of the game – you need to be resilient, you need to be strong enough to continue working despite these problems. We need to be humble, we need to be patient. We need to be honest with the data: when a hypothesis does not hold true we need to drop it and start a new one. It's no good fooling ourselves.' He pauses to reflect on the work of many years past. 'I'm still very proud of what we did.'

Maira could find some consolation in the fact that every other pharma company suffered the same challenges: every pan-PI3K inhibitor and dual-action PI3K/mTOR inhibitor failed. But it was a pretty empty consolation. There had been such great hope of breakthrough cancer treatments based on the PI3K pathway – for patients, for physicians, for the whole medical and pharmaceutical community. Now, almost every scrap of hope was gone.

Almost. Novartis still had one card left to play.

To understand what happened next, we need to go back in time. It's 2006, and Maira and Garcia-Echeverria are full of hope and expectation for their pan-PI3K inhibitors. A young biologist joins the team to assist Maira with his cell assays. Her name is Christine Fritsch and she's never worked in oncology before. In fact, her previous role at Novartis was in ophthalmology. Cancer is quite a culture shock for her.

'I had no real awareness of oncology,' she recalls. In her first years at Novartis, the oncology department remained terra incognita.

And yet oncology was inevitably going to be a central part of her life, whether she liked it or not. 'I always avoided cancer,' she confides. 'I didn't want to face the monster that killed my father.'

She was just eleven years old when her father, a heavy smoker, was diagnosed with lung cancer. 'Cancer impacted me the most devastating way one can think of,' she says. He fell ill just before Christmas, had surgery in February and died a week later. 'I didn't get a chance to see him.'

Fritsch always had an enquiring mind. 'As a little child, my garden was my laboratory. Science was my playground.' She would endlessly study flowers and insects, a young naturalist destined to become a biologist. Her grandfather would take her into the forest every day and introduce her to birds, fruits and mushrooms. 'He was really one of my most important mentors. He taught me to observe.'

She considered jobs in various life sciences before applying for the ophthalmology position at Novartis in 2000. For five years she researched treatments for myopia, studying the biology of the disease and identifying suitable drug targets. Then one day she received a corporate email announcing that the ophthalmology department was closing down.

It was the only job she'd ever had; losing it devastated her. When a colleague mentioned an open role in the Novartis oncology team, she went for it. Her days of avoiding cancer were over.

You may have noticed something is missing from this story. The osimertinib chapter was all about mutant genes – the foundation of cancer biology – yet for the last few pages we've been talking almost exclusively about an enzyme, PI3K. Where, you might be wondering, is the mutant gene?

Victor Velculescu was born in Romania, but he's lived in the United States since childhood. As a doctoral student in molecular biology, when he was just twenty-five, he developed a methodology for measuring individual gene expression across a whole genome. His innovation, SAGE, shows how active each gene is within a particular cell, allowing researchers for example to compare normal cells with tumour cells, and so identify potential oncogenes.[2] SAGE paved the way for many of the high throughput sequencing and genomic technologies we use today.

As a graduate of the Johns Hopkins MD/PhD programme,

Velculescu had intended to practise medicine. 'But I found I was enjoying the science so much,' he says, 'and I could see I would have a greater impact through research than I ever could as a doctor.'

In 2003, he turned his attention to the genes that drive tumour cells. Velculescu had been interested in cancer signalling pathways for some time, and he understood the key role kinases like PI3K play in oncogenesis. He had developed some specialised tools that enabled him to analyse oncogenes, and his Johns Hopkins lab had invented some of the world's leading genomic sequencing and mutation detection technologies, as well as advanced bioinformatics (computer systems used to analyse biological data).

Together with a postdoc, Yardena Samuels, Velculescu deployed these cutting-edge technologies to sequence a range of genes thought to encode components of the PI3K family. They compared the genes in colorectal tumour cells with the same genes in normal cells from the same patients. Even with their resources, this was no easy task at the time. If you compare a healthy genome with a cancer genome, you are likely to see at most one hundred differences in the tens of millions of coding DNA sequences. Of those, perhaps five to fifteen might actually have something to do with the cancer. A gene responsible for cancer amongst all the other genes in a tumour resembles a needle in the proverbial haystack.

The results astonished them. 'We found all these tumours with mutations in a single gene,' says Velculescu. Out of 234 different colorectal tumours, taken from patients with diverse backgrounds and conditions, they found seventy-four with mutations in exactly the same gene.[3]

The gene, *PIK3CA*, was already known: it encodes part of the alpha isoform of PI3K.

They next asked if *PIK3CA* was implicated in other cancers. In addition to colorectal cancers, the researchers found *PIK3CA* mutations in some brain cancers, gastric cancers, breast cancers and lung cancers. This one gene, it seemed, played a role in oncogenesis throughout the body.

But they still couldn't be sure that *PIK3CA* was driving cancer growth. The presence of a mutation in multiple cancer tumours is not proof of causality. So they ran experiments to see what effect the *PIK3CA* mutations actually had in human cells. Velculescu and Samuels measured the level of PI3K activity in a cell line with the normal *PIK3CA* gene. Then they set up the same cell line with the mutant *PIK3CA* gene and showed that PI3K activity was significantly higher. Mutant *PIK3CA* caused more phosphorylation; the PI3K pathway was in overdrive. A causal link between the gene and excessive cell proliferation was clear.

Nearly twenty years after Cantley's discovery of PI3K, Velculescu and Samuels had demonstrated that the gene encoding one of its subunits was commonly mutated in cancer.

The two researchers actually underestimated the importance of *PIK3CA* in that initial study, as Velculescu and colleagues showed a few years later when they completed the first genome-wide sequence analysis of human cancers. We now know that *PIK3CA* is the most mutated gene in breast cancer, with mutations in around forty per cent of tumours, and is one of the most frequently mutated genes in any human cancer. These mutations cause PI3K to signal without regulation – it is permanently switched on.

In the 2004 paper reporting their initial results, Velculescu and Samuels concluded, 'If future experiments verify that mutational activation of *PIK3CA* is essential for tumour growth, specific inhibitors of *PIK3CA* could be developed for treatment of the large number of patients with these mutations.'[4]

For Michel Maira, it was confirmation that PI3K was a worthy target for cancer research. He also understood the significance of this gene's specific role, encoding a subunit of the alpha isoform of PI3K (denoted *PI3Kα*); Velculescu and Samuels' discovery suggested the most effective way to disrupt oncogenesis would be to target PI3Kα, *not* the other isoforms of PI3K.

Maira's three candidate molecules were all pan-PI3K inhibitors: they acted against all isoforms of PI3K. But a more selective inhibitor, targeting only PI3Kα, might be a worthwhile avenue to explore. In medicine, the more selective a drug is, the fewer side effects it typically has. If a molecule blocked PI3Kα while leaving the other PI3K isoforms untouched, then it might be a safer therapy for *PIK3CA*-mutant cancers. In particular, it might reduce the hyperglycaemia risk posed by pan-PI3K inhibitors: block only the alpha isoform and that still leaves three other PI3K isoforms active in the membrane to transport glucose into the cell.

But Maira was already working flat out on the three pan-PI3K inhibitors. 'I didn't have the bandwidth to look after that molecule too,' he explains.

So he set up the alpha-selective project and handed it over to his new colleague, Christine Fritsch.

Fritsch had been working in oncology for less than a year, but she took up the mantle of this 'side project' with enthusiasm. With lead chemist Giorgio Caravatti, she used a set of established assays to screen potential compounds for PI3Kα selectivity. These *in vitro* assays were biochemical: they contained no cells, just purified PI3K proteins mixed with reagents that would reveal the phosphorylation activity of each PI3K isoform. With these biochemical assays, Fritsch's team could test hundreds of different chemical compounds to see if any inhibited PI3Kα activity while leaving the other PI3K isoforms untouched.

Next came the most important step. Fritsch had to show that PI3Kα-selective inhibitors could tackle cancer in tumour cells with a *PIK3CA* mutation. For this, she needed cell-based assays – live tumour cells in a petri dish. She assembled a collection of *PIK3CA*-mutant cancer cell lines from a variety of different cancers. With these, she could assess each candidate inhibitor to see what effect it might have on the PI3K signalling pathway and whether it slowed tumour cell proliferation.

Finally, any successful candidate molecule would need to be tested in animals to demonstrate efficacy and to show they were safer than a pan-PI3K inhibitor. Fritsch and her two *in vivo* pharmacologists created animal models of *PIK3CA*-mutant tumours: human tumour cells with the mutated gene were injected under the skin of mice, where they proliferated to form solid tumours. The mice could then be treated with the candidate molecules at a range of doses to test for both efficacy and side effects.

Now she just needed the right molecules to assess.

The lead chemist, Giorgio Caravatti, had been investigating and targeting kinases since the late 1980s. 'All my life, actually, I worked on kinases, it feels like,' he laughs.

Caravatti's team of chemists focused on three classes of compound. One had been suggested as a possible alpha-selective candidate by Maira and Garcia-Echeverria. Another was identified by a Novartis team researching PI3K inhibitors as possible treatments for respiratory diseases in Horsham, UK. The third was their own selection.

For the first year, Caravatti and Fritsch worked by trial and error, testing different variants of each compound class for alpha selectivity in the *in vitro* assays. They were looking for a molecule that would block the ATP binding site in PI3Kα but not the other isoforms. But, without being able to 'see' the exact shape of the binding site, they were to some extent – like most drug hunters before them – fumbling in the dark.

Then a technological development in a quite different field helped focus the search.

X-ray crystallography has been used in biological research for nearly a century, but only in the last few decades has it become a practical tool in drug discovery. It is a marvellous technique that enables researchers to make 3D images of individual proteins and other molecules. It is a bit like using a super-powerful microscope to see a single molecule.

In 2007, another research group used this technique to visualise and publish the molecular structure of PI3Kα. The Novartis team already had a structural representation of another isoform, PI3Kγ, with an inhibitor locked in its ATP binding site. When combined, these two pieces of graphical information allowed them to model exactly how candidate inhibitors might fit inside the ATP binding site of PI3Kα but not the other isoforms – PI3Kβ, PI3Kγ or PI3Kδ.

Pascal Furet had by now joined Caravatti's team. His particular skill lies in using computer-aided design to model the exact physical shape of a molecule and predict how it will interact with other molecules. With a PhD in theoretical chemistry, he reckons if you want to understand molecular interactions, getting to grips with the physics of a molecule is vital.

Lew Cantley might claim everything in biology is explained by chemistry; Furet would counter that everything in chemistry is explained by physics. Ultimately, drugs work by binding to particular target molecules in the body, and whether that binding happens or not is usually down to the physical shape and energy level of the drug and the target.

With a structural model of the binding site of PI3Kα, Furet was able to suggest theoretical shapes of molecules that might fit inside the site, blocking ATP, without inhibiting any of the other PI3K isoforms. Caravatti's team of chemists could then select real-world molecules that matched Furet's guidance.

The most promising molecule to emerge from this process of rational design was based on the lead from the Horsham, UK team: a potential cancer drug had originated in a research programme targeted at respiratory diseases. I find that an inspiring illustration of the value of pharma companies working across multiple disease areas, and it's a reminder of the importance of cross-fertilisation between scientists in different fields.

In Fritsch's assays, the molecule inhibited PI3Kα but none of the other PI3K isoforms. It slowed tumour cell division *in vitro*. And it showed much stronger anti-proliferation effects in PIK3CA-mutant tumour cells than non-mutant tumour cells, confirming the project's fundamental hypothesis.

The team needed to refine the molecule's pharmacokinetic properties – how it was absorbed and metabolised in the body. In particular, they wanted to extend its half-life – to make it last longer in the bloodstream. Again, rational design and Furet's models proved invaluable: understanding exactly which part of the molecule interacted with the unique amino acids in PI3Kα helped them tweak its structure without diminishing its alpha-selectivity.

By switching out a part of the molecule that was not involved with the PI3K amino acids, they were able to slow the rate at which it was metabolised.

In May 2008, Vito Guagnano, a talented Italian chemist who had joined the team a year earlier, synthesised an alpha-selective PI3K inhibitor that fulfilled all Caravatti's criteria. It was labelled BYL719. Fritsch tested the molecule in the animal models. It showed excellent inhibition of PI3Kα for up to twenty-four hours in mice and rats. The pharmacological properties were encouraging, suggesting good pharmacokinetics in humans.[5]

'Everything was converging on the same conclusion,' Fritsch recalls. BYL719 worked, was selective for the alpha isoform of PI3K, and was found to be most effective against cancers with the PIK3CA mutation. But how would it fare in a human being?

The early human trials were led by Cornelia Quadt. Having grown up in East Germany, Quadt received her medical training in Moscow and then worked as a research physician at a cancer hospital in East Berlin, where she supported drug trials. She was well aware how essential new treatment options were for cancer patients, and was fully familiar with drug development procedures long before she crossed over to the pharmaceutical industry.

Together with physicians in specialised cancer hospitals, Quadt recruited patients with all kinds of solid-tumour cancers, including head and neck, colorectal, lung and breast cancer. The physicians ran genetic tests on tumour biopsies and selected only those patients with a *PIK3CA* mutation.

'That was very novel at the time,' observes Quadt. 'It was pretty groundbreaking.' The imatinib team had done something similar a few years earlier, but not from the very start of human trials. Precision medicine – administering particular drugs to patients with particular genes and other characteristics – was still in its infancy, and Quadt was helping to establish its principles.

The first positive result, when it came, was thrilling. 'We were having a regular call with our investigators, discussing the patients,' recalls Quadt. The physicians dealt with a number of routine matters before one of them delivered the message they'd all been hoping for. 'One said, "OK, with this patient, actually we just did the CT scan and we found a tumour shrinkage."'

The patient had metastatic breast cancer. BYL719 had shrunk the diameter of the metastases – the tumours that had spread around her body – by more than thirty per cent. In clinical terms, it was an 'objective tumour response' – a sign that the molecule might be effective.

Quadt called Fritsch immediately and gave her the good news. 'It was a magic moment,' said Fritsch later. 'I will remember that phone call forever!' There was an additional, professional satisfaction: the mid-level dose that had proved efficacious was within the narrow range that Fritsch had predicted based on her animal models. Further studies found the drug's anticipated side effects to be manageable.

In another positive turn of events, an activated PI3K pathway was implicated in the commonly observed resistance to standard hormonal treatments for breast cancer. BYL719, the team hypothesised, could be really helpful in overcoming that resistance when prescribed in combination with hormone-based medicines. 'That was the best idea we had,' Quadt says with a smile.

By the end of that first phase of human trials, 221 patients had been dosed with BYL719 and the investigators were able to compare responses across different types of cancer. The best results were seen in patients with breast cancer.

And so BYL719 came to be seen primarily as a breast cancer drug,* for use in combination with hormone treatments in patients with a *PIK3CA* mutation.† It was named *alpelisib*.

* * *

* Trials are ongoing to establish whether it can benefit patients with other cancers.

† Once they had established clear efficacy in *PIK3CA*-mutation breast cancers, the trial team went back to look at those patients who lacked the mutation. Might they benefit too? A number of non-mutation patients were enrolled in the trial but the results were as expected: they did not see the same increase in progression-free survival as patients with the *PIK3CA* mutation.

The Phase I trial was so successful that the team could have proceeded straight to Phase III. But drug development is often about difficult choices, and at the time Novartis still believed in the potential of pan-PI3K inhibitors and still prioritised buparlisib. The company preferred to invest its available resources in an extensive set of buparlisib trials covering multiple types of cancer. Although Fritsch, Caravatti and Quadt continued to champion alpelisib, they were unable to obtain funding for the pivotal clinical studies that would bring it closer to FDA approval.

Meanwhile, buparlisib was surging ahead. 'They had a huge study programme – BELLE-2, BELLE-3, BELLE-4 – and plans for other studies,' says Celine Wilke. A former professional athlete, Wilke is a physician with a specialisation in clinical pharmacology. She has dedicated her career to oncology. Seeing the limitations of treatments available to cancer patients, and coming from a family of engineers, she wanted to find better therapies. At the Hannover Medical School in Germany, she helped set up a large clinical trial unit and found she really enjoyed running complex drug trials. Wilke joined Novartis in 2014, in time to take part in the last of the buparlisib trials.

'PI3K was a very hot target at the time,' she recalls. 'I was very excited to join this project.'

But unfortunately, as we know, the high expectations around buparlisib were not realised. The only PI3K card left to play was alpelisib. The Phase III trial began in December that year, with Celine Wilke at the helm.

'Christine Fritsch, Cornelia Quadt and Celine Wilke really kept this going,' says Michelle Miller, who was to take over the alpelisib programme a few years later.

Following its experience with the pan-PI3K inhibitors, Novartis proceeded with just one alpelisib study in a single indication – breast cancer. It was called SOLAR-1. Compared to the more expansive buparlisib trials, which covered a number of different cancers in multiple studies, SOLAR-1 was a modest affair.

SOLAR-1 would test alpelisib in combination with fulvestrant, an established standard-of-care hormone therapy, for the treatment of metastatic breast cancer. In 2015, the study began to enrol post-menopausal women with advanced breast cancer* from all around the world. Over the course of three years, the trial would assess whether patients with *PIK3CA*-mutant breast cancer treated with fulvestrant and alpelisib achieved significantly longer progression-free survival than those on fulvestrant alone.

A moment of drama comes midway through most Phase III clinical trials when an independent panel carries out an interim analysis on the results to date. This panel of experts gets to see the results unblinded. They aren't allowed to reveal the results to the trial team, but they can call a halt to the trial if there is either a significant safety issue or evidence of 'overwhelming efficacy'. A decision to stop at this point is therefore either very bad or very good news. For SOLAR-1, the interim analysis concluded that the trial should continue, indicating that there were no major safety concerns but nor was there yet clear proof that the drug worked.

Anxiety levels rose. 'Everyone was thinking, "OK, this will never be a positive study," ' says Celine Wilke. So much was riding on this last roll of the PI3K dice.

While alpelisib was undergoing the Phase III trial, Roche/Genentech were developing their own PI3K inhibitor, taselisib (*-lisib* is the INN suffix for PI3K inhibitors). Like alpelisib, taselisib targeted mainly the alpha isoform. Like alpelisib, it was in a Phase III trial for treatment of metastatic breast cancer. It was even being combined with the same hormonal therapy, fulvestrant.

Roche's taselisib and Novartis's alpelisib were targeting the same disease with the same mode of action.

* More precisely, hormone receptor-positive, human epidermal growth factor receptor 2-negative (HR+/HER2-) advanced breast cancer.

Imagine, then, the dismay in the alpelisib team when Roche announced in June 2018 that it was scrapping the whole taselisib programme. Progression-free survival improved by only two months when patients were treated with taselisib, and the drug caused multiple, significant side effects. The benefits did not sufficiently outweigh the risks, and taselisib was terminated.

'So we [at Novartis] really thought we also had a high chance of a negative result,' says Wilke. 'It was very challenging at the time to motivate everyone.' Roche's negative outcome, following on from the inconclusive interim analysis and all the earlier failures of the PI3K programme, put everyone in a gloomy mood.

'These PI3K inhibitors were kind of dropping off,' observes Michelle Miller. A paediatric oncologist, Miller had dedicated four years of her life to the failed BEZ235 and buparlisib trials. She took on the leadership of the alpelisib team in August 2018, just in time to oversee the end of the Phase III trial.

The so-called *first interpretable results* were due on 12 August. No one outside the statistics group analysing the data knew what the verdict would be. 'Don't ever play poker with a statistician,' laughs Miller, 'because they share nothing.'

On the big day, Miller was up at 6 a.m., checking her phone. A text message from a team member who'd risen even earlier to scan the report was waiting for her: *We have a drug.*

Alpelisib's efficacy was confirmed in patients with *PIK3CA*-mutant breast cancer; progression-free survival was significantly greater than in patients treated with a placebo.

Celine Wilke was at work when she received the report and she was thrilled to see the positive result. But she had to keep her excitement to herself. Pivotal trial successes in pharmaceutical companies are a very big deal for the financial markets, so the team aren't allowed to talk about their results – even with their colleagues – until a formal press release has been issued. 'We weren't allowed to share [the good news], which is much harder if you sit in open space,' laughs Wilke.

* * *

Alpelisib was approved by the FDA in May 2019.* This came after a nearly twenty-year drug discovery effort, and almost thirty-five years after Lew Cantley's discovery of PI3K. The UK's MHRA approved the drug in December 2021.

Alpelisib is not a cure: it does not keep patients with this advanced form of breast cancer alive indefinitely. Cancers evolve in response to drugs, develop resistance, fight back. Like most other cancer medicines, alpelisib can only keep metastatic cancer at bay for a while.

But it does buy these patients perhaps another eight months of life.[6] For cancer patients I've treated, this might have meant they could attend their child's graduation or see the birth of their first grandchild.

Moreover, by successfully creating, refining and trialling alpelisib, at great cost and effort, Christine Fritsch, Giorgio Caravatti, Cornelia Quadt, Celine Wilke and all their colleagues have resurrected some of the hopes and expectations once invested in the PI3K pathway. After the failure of so many PI3K projects across the industry, their work has allowed us to keep believing. Targeting the PI3K pathway can indeed yield viable cancer medicines.

Without their sustained commitment, this entire area of molecular biology might have been written off by the pharma community as an expensive dead end. Instead, there are now even more PI3K inhibitors, including mutation-specific inhibitors, under development across the industry.

Progress in the development of cancer treatments remains too slow. There is still no cure for the disease that killed Christine Fritsch's father or mine. But the examples of osimertinib and alpelisib buttress our faith: despite all the setbacks, all the failures, all the patients not saved, the struggle is worth it. We're steadily

* In combination with fulvestrant for postmenopausal women, and men, with hormone receptor-positive, HER2-negative, PIK3CA-mutated, advanced or metastatic breast cancer, as detected by an FDA-approved test, following progression on or after an endocrine-based regimen.

advancing. Importantly, we're progressing with a clearer, fuller understanding of the underlying biology of cancer. Greater knowledge will deliver greater impact. More and better cancer treatments are coming.

CHAPTER 6

The Virus That Stopped the World

Condition: Covid-19
Virus: SARS-CoV-2
Innovation: Nirmatrelvir [nir-muh-TRELL-veer]
Company: Pfizer

I BEGAN WRITING THIS BOOK IN the summer of 2020. Covid-19 had just upturned all our lives. My two kids, still in shock at being kicked off their college campus, were holed up in our apartment in Basel, Switzerland, keeping odd hours as they followed a US curriculum on a six-hour time difference. From my makeshift home office, I was steering Roche Pharma Research and Early Development (pRED) through the pandemic, and coordinating our R&D response to the new disease. At the same time, I was learning to find small pleasures in simpler activities – a daily walk along the Rhine River with my wife, digitising old family photographs, playing a socially distanced game of tennis when guidelines allowed.

The hunt for a vaccine was the top priority across the industry, and many of the innovation themes we're uncovering in this book were fundamental to those life-saving discoveries.[1] Meanwhile, an equally urgent search was under way for antivirals to treat

Covid-19, although it was conducted with considerably less media attention. An *antiviral* is a medicine – a compound designed primarily to fight an established viral infection. That makes it quite different from a vaccine, which is given to healthy people to prevent or lessen future infection.

One of the most exciting Covid-19 antiviral breakthroughs was achieved by a team at Pfizer, where I went on to serve as Chief Development Officer. I arrived some months after its discovery, and played a small part in its late development.

It's a story that unfolds at breakneck speed in very trying circumstances. But it begins with a simple speculation, back in December 2019, when the term Covid had not yet been coined and most people had only a faint memory, if any, of a related disease: *Severe Acute Respiratory Syndrome* (SARS).

Fortunately, Jennifer Hammond, then a medical director in Pfizer's Anti-Infectives group, was paying attention to the international news. She heard about the first reported cases of a new SARS-like disease in Wuhan, China, and they brought to mind a SARS antiviral programme from early in her career. Sixteen years previously, Pfizer had come up with an experimental molecule that showed promising pre-clinical activity against SARS. Might it work against this emerging threat to public health?

Hammond grew up in Washington state, on the West Coast. From an early age, she was fascinated by infectious diseases. 'I used to go to the library and check out books on the Black Death.' Unusual reading choices for a child, but Hammond was clearly hooked on pathogens. Ignaz Semmelweis, who showed that doctors could reduce deaths in childbirth simply by washing their hands before delivery, was an inspiration.

The budding biologist spent her summers working on the family's potato farm. Hammond's father assigned her jobs that would further develop her interest in science, such as collecting samples from the fields, taking them to a lab to be analysed, and plotting the results on charts used to guide fertiliser application and harvesting. At college and graduate school, Hammond pursued her interest in infectious diseases. For her PhD, she investigated HIV resistance to antivirals, the perfect springboard to a job in Pfizer's Antiviral Discovery Group in 2001.

The group had a particular focus on *proteases*, enzymes used by viruses to cut large proteins into smaller parts that then enable viral replication. Imagine a string of sausages that need to be separated before they can be cooked. A typical virus has a polyprotein – a string of proteins – that needs to be cut up, liberating the proteins to play their part in making new copies of the virus. The protease is like the scissors used to cut up the string of sausages: it cleaves – and thus activates – the viral proteins.

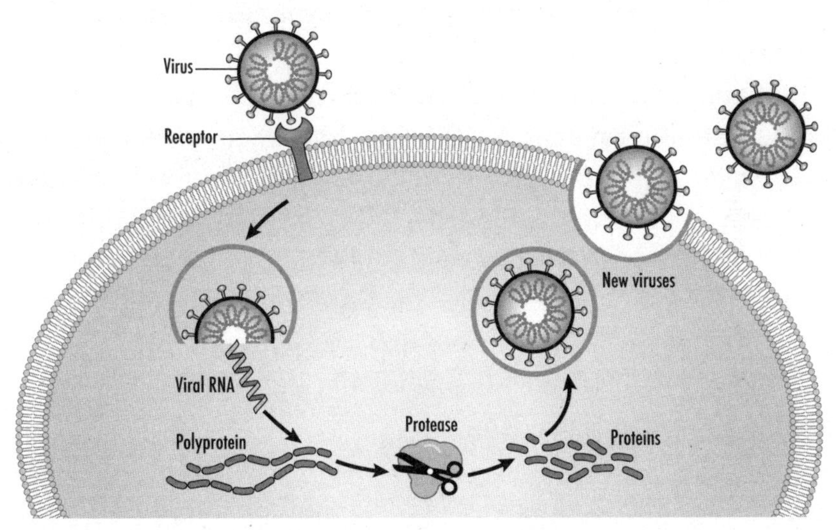

Figure 17. The role of a protease in a viral life cycle.

Many viruses rely on proteases to complete their life cycle, including HIV, hepatitis C and rhinovirus (a common cold virus). Proteases are great targets for antiviral drugs because they are so critical to viral replication, and because these viral proteins don't exist in humans. *Protease inhibitors* have proved their worth as safe and effective therapies for HIV and hepatitis C. An inhibitor that binds tightly to a protease is likely to remain effective even if the virus begins to mutate. The famous spike protein on the surface of the Covid-19 virus mutates rapidly, making it challenging for vaccines or antivirals that bind to it to keep up. By contrast, the active binding site on the *main protease* (called 'Mpro' or '3CL protease') rarely changes.

In the early 2000s, Hammond's team was primarily focused on testing new HIV protease inhibitors. Other teams within the Antiviral Discovery Group worked on novel inhibitors of hepatitis C virus and rhinovirus. Much of their work was conducted in a restricted access area Biosafety Level 3 (BSL-3) lab, equipped with a negative-pressure antechamber, double gloves, goggles and high-spec air filters. When SARS, a fast-spreading disease caused by a new coronavirus, emerged in 2003, the Antiviral Discovery Group was quick to redeploy resources and expertise in the hunt for protease inhibitors.

Although Hammond never worked on SARS, she had a front-row seat to the whole endeavour. The group was small and tight, and every other week they shared progress and ideas across the different teams. The SARS effort was intense, a high-priority mission to address what seemed at the time to be an unfolding health emergency. I remember the SARS outbreak well: I was a practising physician at the time, and we were concerned that the troubling news of high fever, pneumonia and death in Asia might lead to a worldwide pandemic.

The SARS team played their part heroically and within a few months had identified a range of compounds that seemed to work as inhibitors against the SARS main protease. The most promising molecule was labelled PF-00835231, or *5231* for short.

The molecule was undergoing pre-clinical studies when the epidemic burnt itself out. In the end, just over eight thousand people are thought to have become sick with SARS, and 774 died.[2] The disappearance of SARS was great news for humanity, but it meant there were no patients to recruit for clinical trials of Pfizer's new investigational drug.

The team shared the discovery of 5231 with the broader scientific community, and then left the molecule on a metaphorical shelf.

Institutional memory is immensely important in many areas of life; in scientific innovation, it can be pivotal. Though Pfizer's antiviral research programme had been discontinued, and many people had retired or moved on since 2003, Jennifer Hammond was still with the company all those years later, and she remembered 5231.

In January 2020, as it became clear that the new SARS-like virus spreading in China was likely to become a serious international health concern, Hammond pulled together all the information she could find on the 2003 programme and brought it to the attention of one of Pfizer's chief scientific officers, Annaliesa Anderson.

A British microbiologist, Anderson had built her career on bacteria, designing and developing vaccines over two decades at Merck, Wyeth and Pfizer. Developing antivirals was not her area of expertise. Nevertheless, she grasped immediately the potential importance of an antiviral that might work against this frightening new virus, soon to be named *SARS-CoV-2* (Severe Acute Respiratory Syndrome CoronaVirus-2).

As a child growing up in Nottinghamshire, Anderson had initially planned a career in medicine. She attended a taster course on how to be a doctor at the age of sixteen, but quickly realised general practice medicine would not satisfy her scientific curiosity. 'I've always been interested in how things work, particularly small things,' she says. Microbiology allowed her to explore a world of highly consequential organisms that most of us never see.

After talking with Hammond, Anderson began thinking through a possible response. The main problem was that Pfizer was no longer in the antivirals business. Like many of its peers, Pfizer had decided to concentrate its research on other areas – cancer, cardiovascular and immunology diseases among others. The antivirals researchers had retired or been reassigned, and the

BSL-3 laboratories – essential to any research with dangerous pathogens – had been dismantled.

With no formal research unit or established budget for antiviral therapeutic work, Anderson would have to borrow scientists and funds from other units. As chief scientific officer for bacterial vaccines, she could call on support from Pfizer's Vaccines R&D organisation; the knowledge and experience of its scientists would be invaluable. The Hospital unit, her other responsibility, could spare some budget to get the work going. Wuhan was now under quarantine, with at least four people known to have died, and the first case of the 'novel coronavirus' had just been identified in the United States. It was time to act.

In early February 2020, Anderson invited Hammond to New York to present the data on 5231 to the Hospital unit leadership team. It would turn out to be her last work trip before lockdown. The new disease had still not been given a name, but the World Health Organization had just declared a global health emergency. Governments were imposing restrictions on international travel. Against that bleak backdrop, Hammond and Anderson recommended testing Pfizer's 5231 on the new virus.

Anderson also consulted Mikael Dolsten, Pfizer's long-standing head of R&D, who connected her to Charlotte Allerton, the head of Medicine Design. Anderson had met Allerton, another Brit, a couple of times previously, but they had never worked together. Now, she got in touch with the chemist and raised the subject of 5231. 'Oh yes, we've been looking at that,' said Allerton brightly. Institutional memory was alive and well across several parts of Pfizer.

There were three immediate questions for Allerton and Anderson to address. The first was a theoretical assessment: was the new viral target similar enough to the old viral target against which 5231 had been shown to work? Next, there was the biochemical analysis: did 5231 lock on to the target in a test tube? The final question was the biological proof: would 5231 stop the virus replicating inside a living cell?

The two leaders assembled a team and got to work. They established biochemical assays to determine whether 5231 recognised the SARS-CoV-2 main protease, and contacted partner organisations with BSL-3 labs who could help test the molecule against a live virus.

The theoretical assessment gave reason for hope. A team of researchers in China had identified the novel coronavirus and sequenced its genome; they made this information public in January. The main protease was almost identical to that of the original SARS virus (which we now call SARS-CoV-1). There was no guarantee, but the structural similarity between the two targets made it highly likely that 5231 would also work to inhibit the new virus's protease.

The biochemical analysis, too, was encouraging. Allerton's team showed that 5231 would indeed bind tightly to the SARS-CoV-2 main protease, potentially inhibiting its enzymatic action. The business end of the SARS-CoV-2 protease is a *cysteine residue* (cysteine is an amino acid). We can think of that cysteine residue as the blades of the molecular scissors the virus uses to chop up its polyprotein, releasing the constituent proteins that are essential to its replication. 5231 binds to the cysteine residue and gums up those blades. Picture a toffee stuck between the blades of your scissors and then imagine trying to cut up that string of sausages.

Now it was up to Anderson's group to pursue the biological proof. Alongside 5231, they decided to investigate a whole range of other potential antivirals. 'We went through our freezers,' says Hammond. The team pulled out dozens of molecules from Pfizer's long history of research and tested them all against SARS-CoV-2.

Their analysis was unambiguous: the most potent by far was 5231.

In this post-pandemic crisis era, it's easy to forget the sense of impending catastrophe we all faced in early 2020. Late February saw hundreds of people on a cruise ship in Japan testing positive for the virus. The ship was quarantined and passengers started dying. A fortnight later, the plight of cruise ships was eclipsed by a maelstrom engulfing Italy. Hospitals there were overwhelmed by the sudden influx of very sick people; hundreds of deaths were reported. Italy went into lockdown. The virus's swift spread sparked panic: no one knew exactly *how* it was transmitted. We were scared we could be killed by our neighbour, by a fellow passenger on a plane, by a stranger in the supermarket or on public transportation – just by breathing.

The new disease was named *Covid-19* and was now the lead story across the world.

Pfizer had publicly shared its work on SARS years before, but as the threat escalated, the company released further data on 5231 into the public domain. Every pharma company and research institution now had the opportunity to use Pfizer's data to build a SARS-CoV-2 protease inhibitor.

Pfizer was not alone in this new openness. When Covid-19 hit, pharma companies, academia, regulators and public health authorities came together in a remarkable way to share scientific and medical ideas, information and resources that might contribute to the fight against this deadly new disease.

As heads of early R&D at our respective companies at the time, Mikael Dolsten and I were asked to join ACTIV (Accelerating Covid-19 Therapeutic Interventions and Vaccines), a public-private partnership formed to develop a coordinated research strategy to speed development of the most promising potential treatments and vaccines. ACTIV brought NIH together monthly with its sibling agencies, including the Biomedical Advanced Research and Development Authority, Centers for Disease Control and Prevention, and the FDA. Other government agencies such as the Department of Defense and Department of Veterans Affairs were also represented. Operation Warp Speed, the European Medicines Agency, and representatives from academia, philanthropic organisations, and numerous biopharmaceutical companies all joined the partnership.[3]

This level of collaboration really was unprecedented.

* * *

To conduct any further development work, a sample of 5231 was needed. It might have taken weeks or months to synthesise a new batch: unusual starting materials were needed, requiring around ten steps to formulate, with each step taking three to eight days to run. These would need to be sourced from specialist chemicals firms around the world. Putting everything together, with all the appropriate batch checks, purifications and regulatory records, could be a long and complicated business.

Luckily, Pfizer still had – preserved since 2003 – a small batch of 5231. It was just a few grams, but it was enough to get started. They had to share it out in milligrams to all the teams that would need to take a look at it – for biochemical assays, biological assays, toxicology tests, formulation – so every gram counted.

When Allerton and her team took a closer look at 5231, they found three significant drawbacks.

The most obvious problem was that 5231 would not work as an oral drug: it could not be taken as a tablet because the molecule would not pass through the gut wall. It could only be administered intravenously. That was all very well for the sickest patients in hospital, but it would make the drug useless for the much wider at-home population.

Then there was the matter of solubility. 5231 did not dissolve well in fluid, making its value as an IV drug questionable.

Finally, there was the stability challenge. 5231 did not stay 5231 for long in the blood. Its half-life was less than one hour: within an hour of administration, half of the drug would already be broken down. That meant 5231 would have to be administered as a continuous infusion. Patients would need to remain on an IV drip for days.

The scenes of fear and chaos in overstretched hospitals around the world underscored how much better it would be to treat people earlier, before they reached the stage of hospitalisation. An IV drug would not be practical in most cases.

For Allerton, Anderson and Dolsten, the mission was clear: draw on the structure of 5231 to design a new, stable molecule that was effective against SARS-CoV-2 and *orally bioavailable* – it could be taken as a tablet. And it had to be done in record time, without compromising safety or quality.

'You saw the images on TV,' says Allerton. 'You can't sit there and not do anything about it.'

5231 was insoluble, unstable and wouldn't pass through the gut wall. Luckily, pharmacokinetics (PK) and drug metabolism were very much in Allerton's wheelhouse. She'd been fascinated by how medicines work ever since, aged ten, she was given antihistamines for her allergies. An outstanding teacher in Suffolk had nurtured in her a love for organic chemistry and an intense curiosity as to how atoms could be arranged in molecules to influence human health. After working as a medicinal chemist at Pfizer's Sandwich site, she moved into the pharmacokinetics and drug metabolism field, focusing on how to get a drug where it's needed in the body.

Hundreds, perhaps thousands, of otherwise valuable molecules have failed in the clinic because of poor pharmacokinetics. Allerton was part of a new wave of chemists who were determined to get to the bottom of the PK problem.

To lead the effort to design a new molecule, Allerton picked Dafydd Owen. Growing up in Newbury in the UK, Owen wanted to be a doctor. He was definitively put off by a physicians' demonstration day. 'I saw the medics with their surgery tools there, and that scared the heck out of me,' he says. In preparation for medicine he'd studied chemistry, and that paved the way for an alternative career. Organic synthesis – making molecules – fascinated him. He took a job at Pfizer's Sandwich site and there learned to, as he puts it, 'perturb biology by design'. It was a formative experience.

'I was just in awe, working in the [hazardous chemical] fume hood next to the guy that made amlodipine* for the first time – a drug that kept my grandmother alive until she was ninety-three.'

* A calcium channel blocker used to manage hypertension.

For Owen, chemistry lies at the core of drug discovery. 'I'm particularly passionate about medicinal chemistry because you *create*. Ultimately every other discipline measures something about your molecule – its toxicity, its potency, its efficacy – but only the chemist can then *do* something about it and invent a better molecule.'

Allerton asked Owen to lead the oral Covid-19 therapeutics team on Friday 13 March 2020, the day Pfizer sent almost all its staff home indefinitely. More than four thousand people had died of the disease worldwide, and cases were now popping up all over the US. The WHO had just declared a pandemic.

The first step for Owen and his team was to go back to the 2003 work and screen all the potential SARS compounds the Antiviral Discovery Group had come up with, in case an orally bioavailable compound was hiding in plain sight. Nothing else turned up in that initial screen: 5231 remained the best candidate.

Their next move was to try to make an orally available *prodrug* of 5231 – a modification of the molecule that would be reversed once it had sneaked past the gut wall. Swallowed as a tablet, the hypothetical prodrug would be absorbed from the gut and then converted into 5231 in the bloodstream. If it worked, it would be a quick and convenient fix.

Scrutinising the structure of the molecule, Owen could see an obvious reason why it wasn't able to cross the gut wall. It had a *hydroxy group* (-OH) and four other *hydrogen bond donors*.* These polarised groups tend to repel the kinds of fatty molecules that make up a biological membrane.

Owen assembled a crack team of chemists via video conference, including lead design chemist Jamie Tuttle, medicinal chemist Matt Sammons, computational modeller Joy Yang, and synthetic chemistry lab leader Matt Reese. The answer, they reckoned, was to 'hide' the problematic groups temporarily by attaching

* Essentially, a hydrogen bond donor is a hydrogen atom bonded to something other than a carbon atom (e.g. O-H or N-H).

something else to them. 'You put a glove on this side of the molecule,' jokes Reese, 'and a hat on that side to cover up the parts that keep it from being absorbed.'

The hydroxy group could be converted, with a small molecular addition, into an ester. Esters are the volatile compounds that give the pleasant smells to fruits and perfumes; they pass easily through biological membranes. Esterases (enzymes in the blood that break down esters) would quickly remove the added group once the disguised 5231 had passed through the gut wall, restoring the antiviral. Further cleavable molecular groups, such as phosphates, could be used to hide the other hydrogen bond donors. Together they would 'grease up the molecule', in Reese's words, rendering it better able to slip through the gut wall.

Working alongside the chemists was a team of drug metabolism experts led by Amit Kalgutkar. Their job was to test the chemists' candidate molecules for the key metabolic attributes they needed: stability in the gut and blood, and ability to cross the gut wall.

Kalgutkar's assays included samples of human gut media – the stuff in our stomach and intestines. They showed that the chemists' prodrug candidates were not going to work. 'They were all hydrolysed,' Kalgutkar explains, meaning they were broken down by enzymes and acids in the gut media. Not only were some of the hats and gloves unstable, the very backbone of the molecule was breaking down in those highly digestive conditions.

It was not an unexpected result. 'Peptido-mimetics all suffer from proteolytic cleavage,' says Kalgutkar. To translate, drugs that try to look like proteins tend to get destroyed by enzymes that have evolved to cut up proteins.

The quick fix was off the table.

So that left just one option: design an entirely new molecule with the antiviral properties of 5231 but none of its pharmacokinetic drawbacks. Owen's team of medicinal chemists would have to do something that normally took years in a matter of months. And they'd have to do it during a global health emergency.

'I was scared of catching Covid,' admits Owen. 'I was disinfecting groceries and leaving them in quarantine in the garage for forty-eight hours. I bought a second freezer because I thought we were going to have to stockpile food. You just didn't know what was going to happen.' His children were old enough to be mostly independent, but he worried about his parents in the UK. 'I told them if they got sick, I wouldn't be able to come.'

'Like the rest of the world, we were grappling with a lot of personal challenges,' adds Allerton.

More and more countries were going into lockdown. By mid-April, over 100,000 people worldwide had died of Covid-19. The British prime minister was in intensive care. New York City, where Pfizer is headquartered, had become the new epicentre of the disease. On a single day, 7 April, 815 NYC residents died.[4] The federal government deployed over a thousand medical military personnel to the city, and the National Guard assumed the sombre duty of collecting the bodies of the deceased.

Speaking to my former academic colleagues who were on the front line in New York hospitals, I heard about the daily challenges and tragedies hospital staff were facing. I was increasingly anxious about my eighty-year-old in-laws living in Queens, while my wife and I were stuck nearly four thousand miles away. How would they keep themselves safe?

Confined to their homes like so many of us, Owen's team of chemists gathered by video call to review the molecular structure of 5231 and decide which bits to keep and which to discard. 'Our task was really to perform molecular surgery,' says Owen.

Their first focus was those parts of the molecule that impeded its progress across the gut wall. They wanted to eliminate the hydroxy group, reduce the number of hydrogen bond donors and increase the carbon content, rendering it more *lipophilic* (fat-friendly). They realised they would have to replace most of the molecule.

But in making these substantial changes, they had to preserve the part of 5231 that locked on to the cysteine residue of the main protease.

With all these requirements in mind, Tuttle and his team got to work designing theoretical molecules that might both inhibit the main protease *and* be taken as a tablet. They came up with many more possibilities than the team could realistically attempt to synthesise, so then it was up to Yang to model how each theoretical molecule might behave in a human body – how easily it would pass through the gut wall, and how well it would lock on to the cysteine residue. Only those molecules that seemed most promising would be synthesised.

We witnessed the power of computational modelling in Chapter 2. During the pandemic, when everyone was trying to minimise time spent in communal spaces like labs, it was invaluable.

Once Tuttle and Yang had done their theoretical work, Reese and his team of synthetic chemists ventured out to make the most promising molecules. 'While I was at home disinfecting my groceries and setting up my home office,' says Owen, 'those brave people went back into the lab.' They needed a special letter from the governor of Connecticut allowing them to travel to work. Pfizer's Groton site is designed for three thousand people, but during much of the pandemic there were only about forty scientists parking in its vast car parks. Reese and his team found themselves walking the echoing halls alone.

The team's basic approach was to create a molecule that mimicked what the SARS-CoV-2 main protease was expecting to 'see'. Instead of regular peptides that are quickly broken down in the gut, they would use artificial cousins that are more resistant to digestive enzymes and therefore more stable. The viral protease would normally bind to a glutamine residue and a leucine residue on the polyprotein (glutamine and leucine are amino acids). So Tuttle and his design team searched for chemical plug-ins that 'looked like' glutamine and leucine. Ideally, these mimics would bind more strongly to the protease than the original amino acids, so the drug would 'out-compete' the polyprotein and gum up the molecular scissors.

The plug-in they chose to mimic leucine was a *bicyclic proline analogue* (a modification of a natural amino acid). Yang had computationally modelled it and found it to be a very good mimetic of what the protease was expecting to see. Moreover, when part of a larger molecule, it was not susceptible to digestive enzymes and so helped with oral bioavailability.

The bicyclic proline analogue does not occur naturally, but it can be synthesised through a long and complicated process that starts with a compound derived from chrysanthemums. The process is expensive and can take months to execute. With no large-scale industrial application at the time, very little of the stuff actually existed. For the initial synthesis of candidate molecules, Pfizer was ordering supplies in milligrams.

As the team zeroed in on a structure that seemed promising, the importance of the rare bicyclic proline analogue became increasingly apparent. If they were to proceed any further in testing – and ultimately trialling – their candidate molecule, they would need a lot more of it. Pfizer's external sourcing team started scouring the inventories of the world's chemical manufacturers. At the time, they reckoned, there were just nine kilograms of the precious substance in existence anywhere on the planet.

Pfizer bought virtually all of it.

Meanwhile, Anderson was swiftly rebuilding Pfizer's antiviral capability, hiring highly experienced virologists and setting up BSL-3 laboratories so that they could safely handle SARS-CoV-2 and be ready to test any potential clinical candidates. One of her team's most important tasks was to develop new *in vitro* assays. Assays used in pre-clinical drug discovery are usually simpler and less stringently calibrated than those needed later to prove a drug's worth for regulatory submission. Anderson took the decision to make all Pfizer's new assays regulatory-quality: it meant more work and expense up front, but would save time getting any successful drug to patients. They were able to speed up the process by collaborating with external partners who could test compounds in high throughput assays.

The dire situation in New York continued through April and into May. The city's Covid-19 death toll reached five figures. The new head of the Pfizer Hospital unit, Angela Lukin, lived in Manhattan at the time, and when I speak to her in 2022 she remembers how the normally bustling streets came to resemble something out of an apocalyptic movie. All the restaurants, theatres and stores that make New York such a vibrant city were shuttered. The schools were closed. Ambulance sirens wailed day and night. Refrigerated trucks were used to store bodies outside NYC hospitals.

'It was so sad and very scary,' Lukin says. 'I remember seeing police cars on every other corner and hardly anyone in the street. The only time I left my apartment was to buy food. Every day at 7 p.m., I would open my window and join other NYC residents in clapping to salute the city's medical workers.'

Lukin was inspired to join the pharma industry after witnessing her mother grapple with immense pain, the result of three types of arthritis for which there were no effective treatments. A breakthrough finally resulted in the first biologic medicine for rheumatoid arthritis; her mother's relief following this milestone achievement became a source of constant inspiration. 'It gave me a sense of purpose, to be able to bring medicines to market that could make a difference in patients' lives.'

Like all of her team, Lukin was incredibly busy in the first months of the pandemic, trying to ramp up supply of essential medicines to hospitals around the world. Demand for some products, such as antibiotics and drugs needed to intubate or sedate patients, was up 600 per cent in some cases. Her team even had to figure out how to deliver supplies to the Navy's hospital ship moored off Manhattan. Every hour of every day was spent triaging supply and coordinating with hospitals globally to try to meet their spiralling needs.

'Lives hung in the balance,' says Lukin. 'Our colleagues were working day and night to do whatever was possible . . . even things we thought would be impossible. At the height of the pandemic, we estimated that one in five hospitalised patients around the world was touched by a Pfizer hospital medicine.'

Lukin did this all while in isolation in her New York apartment. It would be two years before she actually met her scientific counterpart, Annaliesa Anderson, and many other Hospital unit colleagues face to face.

The Hospital unit assessed many other potential antivirals submitted to Pfizer by external researchers. They had the resources to develop new medicines fast, so it was important to investigate every lead. But ultimately, they declined to take on anyone else's molecules – none appeared to be as promising as their own team's discoveries.

Matt Reese finds chemistry innately appealing. 'I love the puzzles,' he explains. 'I love the challenge of figuring out how to ask questions and then trying to use chemistry to answer those questions.' After university, he'd chosen to work just a couple of years in the pharma industry as a prelude to medical school, but then found he didn't want to leave. 'Ultimately, I decided to trade in the chance to help individuals every day for the admittedly risky opportunity to have an impact on a big patient population in need.' Risky is right: by 2020, not one of the many molecules Reese had worked on in his twenty-five years at Pfizer had made it to late-stage human trials. You might be surprised, but this is not at all unusual in our industry, and just goes to show how difficult drug development is.

Each molecule Reese and his team built was tested against SARS-CoV-2, and its drug metabolism attributes were investigated. Amit Kalgutkar's team had created new *in vitro* assays that mimicked the hydrolysis action of gut enzymes, so they could quickly test the stability of each iteration. To investigate oral bioavailability, the team couriered tiny samples of each compound to a contractor who administered it to rats both orally and intravenously, comparing the plasma concentrations achieved via the two routes.

These data were brought together by computational modeller Britton Boras to predict how each molecule would behave in human bodies. Drawing on Pfizer's previous work in protease inhibitors, as well as data from influenza pathways (to reflect SARS-CoV-2's likely infection pattern) and the *in vivo* rat data, Boras built models to assess the likely performance of the new molecules. His findings helped the chemistry team iterate their designs.

Life in the lab was as unsettled by Covid-19 as everything else. The chemists weren't allowed to gather for meetings, so Reese spent much of his time coordinating his team members and keeping them updated. They had to balance two opposing safety concerns: they were mandated to work far enough apart to prevent contagion, while at the same time staying close enough to a colleague to meet lab safety regulations. 'You need to be within shouting distance in case something happens,' explains Melissa Avery, a member of the chemistry team.

For Avery, meeting that rule could be a challenge. One of the youngest team members, she would frequently want to work late or come into the lab at the weekend, when the others had childcare commitments. They had to set up a buddy system so that no one would ever find themselves alone in the lab.

Medicinal chemistry is, for Avery, a perfect job, although not without its frustrations: 'We can spend weeks to months really getting into the fine details of a specific type of chemistry to make a compound that we really want . . . and then find out that we don't want that compound any more.' The hard truth is that most molecules never go anywhere – scuppered by pharmacokinetics, toxicity or market forces. Chemists, like other innovators, have to be willing to pivot regularly.

But with the global Covid-19 death toll passing 300,000 in May, the team all understood the crucial importance of their work. 'It felt like the weight of the world was on our shoulders,' says Reese.

'The time pressure was immense,' agrees Kalgutkar. Still, there was an upside to the sense of mission. 'I have never ever seen a bunch of such enthusiastic people,' he adds.

Work–life balance was a distant memory. The scientists regularly worked through weekends. 'Knowing when to stop each day was a challenge,' admits Avery. Kalgutkar, who worked entirely from home, recalled scheduling meetings at ten o'clock at night when important decisions had to be made.

Both Reese and Avery were glad to be back in the lab. When everyone was sent home in March, Reese found those first few days challenging. His wife was teaching from home, and their two kids were attending school remotely. 'It was definitely chaotic because we had three people trying to navigate three different classrooms in a house that normally sat empty during the day. I think we all benefited from the extra space afforded from me having someplace else to go!'

Being at home posed different challenges for Avery. Stuck in her living room, she felt unable to make an impact. When the call for volunteers to go back on site came, 'I jumped at the opportunity.'

By June 2020, less than three months in, they had a molecule. That is extraordinarily fast. It was designated PF-07314688, or *4688*. The team tested it against the SARS-CoV-2 main protease in biochemical assays and found it successfully bound to the enzyme. It reduced viral replication in cells, seemingly without harming the cells themselves. In rats and other animal species, it was successfully absorbed through the gut wall, confirming oral bioavailability. Anderson and Allerton began laying the groundwork for clinical trials.

For studies in humans they would need a lot more 4688, and that meant using up the bicyclic proline analogue. The manufacturing team were ready to commit all nine kilos of the precious chemical to make as much 4688 as possible. But at that point, Allerton's team received new information that made them hesitate. The chemists had come up with an alternative molecule, designated PF-07321332 – or *1332*. Assay results indicated it was at least as promising as 4688.

By now, the machinery of pre-clinical development had swung into gear, and 4688 was six weeks further down the track than 1332. It would be 4688 that was put in trials first. But the 1332 results made Allerton's team think again about that precious supply of the bicyclic proline analogue. The chemical building block was also needed to make 1332. If they blew it all on 4688, there would be none left to progress 1332.

At the last moment, they called the manufacturing team and asked them to hold back one kilogram of the bicyclic proline analogue.

A few weeks later, Dafydd Owen's phone rang. Early.

'You don't want to get an 8 a.m. call from your drug safety rep,' he said later. It's never good news. The preliminary toxicology studies for 4688 had found a problem: the molecule had an unwelcome effect on red blood cells in pre-clinical animal species; any possibility of a similar effect in humans was an unacceptable risk. 'It was game over for that molecule.'

For the whole team it was incredibly disappointing. They had thought they held the answer to the world's most urgent health problem. Now it was gone.

But there was still 1332. Less advanced, but just as promising in its efficacy, the backup molecule was now the lead candidate. And thanks to that last-minute decision, they still had a kilogram of the bicyclic proline analogue left to make it.

Only about one quarter of 5231's structure is preserved in 1332. Owen's team of medicinal chemists had worked incrementally to improve the potency, stability and oral bioavailability of the original molecule, but the ultimate result was a radically different design. This is sometimes the way in innovation: the path to a completely new concept may be a series of incremental steps. Careful, systematic engineering can yield revolutionary change.

One of the most distinctive features of 1332 is a molecular group called trifluoroacetamide. 'It's the most eye-catching piece of the compound,' says Reese. The group was originally introduced only to test a hypothesis about the part of the molecule that was driving antiviral activity. It was an unconventional group to try – it's usually not very stable. But the team found it provided a remarkable improvement in gut wall permeability. They kept testing its stability and it never broke, so it stayed in the design.

Usually, a pharma company will take a step-wise approach and wait for toxicology results before incurring all the costs of planning and setting up clinical trials. But with the global death toll passing one million, Pfizer took the bold decision to push ahead with 1332. They would proceed 'at risk', investing the money to prepare for clinical trials, even as they awaited the pre-clinical results that would determine whether or not they could test the drugs in humans. The call went out to specialist chemical providers around the world to start manufacturing the bicyclic proline analogue in much larger quantities.

The last months of 2020 were a particularly grim period for much of the world, with renewed lockdowns blighting seasonal festivities and a further acceleration in the spread of the disease. Cumulative global deaths rose from one million in September 2020 to two million in January 2021. One bright spot in that dismal winter was the announcement of emergency use authorisations and conditional approvals for the Pfizer-BioNTech and Moderna vaccines. These breakthroughs were achieved in record time and represented a kind of salvation for humanity. The most vulnerable people and front-line health workers were swiftly vaccinated in those countries with access to the first limited supplies of the vaccines. Further vaccines from Janssen and Oxford-AstraZeneca would soon follow.

But even with the vaccines, an antiviral was desperately needed. Many people would not be able to be vaccinated, and many more would become infected and fall seriously ill despite vaccination, due to variability in immune responses.

The toxicology team worked through the holiday season to complete all the checks needed before 1332 could be tested in humans. In early January 2021, they gave the all-clear. 'It was a great moment,' Owen remembers. With all the documentation already in place, the FDA moved as swiftly as possible to review it and subsequently granted approval for human trials.

Sandeep Menon was in charge of early clinical development. A physician in Mumbai, India, before he moved to the US for his doctoral studies, Menon chose to specialise in biomathematics. With his team, he drew on this expertise to design a very innovative '5-in-1' adaptive Phase I study that would compress all the usual tests (single ascending dose, multiple ascending dose, bioavailability assessment etc.) made in healthy volunteers from the standard six months to just six weeks.

Their data would be analysed by Art Bergman, a clinical pharmacologist who was responsible for assessing whether the drug worked and what effect it had in humans. Bergman had found the pandemic much harder than some of his friends and colleagues: 'It was really difficult because I thought about Covid twenty-four hours a day. I'm sure a lot of people did, but I'd go to work and I'd do Covid, and then I'd stay home because of Covid, and then talk to my friends about Covid, so it was hard to escape.' Domestic life wasn't easy either, with three boys stuck at home and struggling with school via Zoom. 'There was a good deal of unhappiness in the house,' he recalls, 'but I was very busy and had to juggle that.'

The big question for this first study was whether they could achieve a sufficiently high concentration of 1332 in the human bloodstream to be effective against the virus. The molecule was somewhat more stable than 5231, but it too had a short half-life. How big would the dose have to be to yield an effective drug concentration? 'The PK [pharmacokinetics] was the key point,' says Bergman.

Menon's team recruited two cohorts of eight healthy volunteers in New Haven, Connecticut, just down the road from the Groton labs. In each cohort, six people were given doses of 1332; the other two were given a placebo. Blood samples were taken a day after administration to see what had happened to the drug. Bergman and his colleagues set up a rapid overnight analysis system at Groton so they would have the PK results analysed early the following morning, enabling a timely decision on the next dose to test.

As always, they started with a low dose and raised it sequentially. But where normally the clinicians would have to wait a week before raising the dose, thanks to the trial design and the fast turnaround of results, they were able to administer and assess two dose increases per week.

The results were encouraging. There were no significant safety issues observed: the volunteers tolerated the drug well at the assessed doses. Critically, 1332 was absorbed well through the gut wall into the bloodstream. It was orally bioavailable – a good candidate for a tablet.

The team wanted to achieve very high, safe concentrations of 1332 in the blood, to have the best chance of stopping viral replication and preventing patients from becoming seriously ill. High concentrations of antivirals also reduce the risk of drug resistance emerging over time. Given the short half-life of 1332, this was going to be a challenge. The initial human study confirmed the problem: the molecule was metabolised – broken down – too quickly. So to achieve the necessary concentrations of 1332 in the blood, the team decided to co-opt another drug, called *ritonavir*.

All drugs are eventually metabolised and eliminated from the body. The question is *how fast* they are metabolised and by what mechanism. Like a lot of drugs, 1332 is metabolised primarily by an enzyme called cytochrome P450 3A4 (CYP3A4) in the liver and intestine. This enzyme breaks the molecule down into inactive fragments that can be excreted.

Ritonavir is a medicine that was originally approved in the 1990s for use against HIV. However, it also blocks the action of CYP3A4, interrupting the primary metabolism pathway for drugs like 1332. So if 1332 was co-administered with a low dose of ritonavir, it should last much longer in the blood.* Ritonavir had already been successfully used in this way to boost concentrations of other antivirals.†

Allerton and her colleagues had long planned for this possibility: supplies of ritonavir had been secured, and the drug was already written into the clinical trial protocols as an option for boosting 1332 concentrations. This careful planning and forward thinking saved weeks of development time, allowing the two drugs to be tested in combination in the next human study.

The results were exactly as hoped. Twenty-four hours after the first dose was co-administered with ritonavir, the concentration of 1332 in volunteers' blood remained above the level needed to kill the virus.

The team was becoming more and more confident, but they still had no clinical efficacy data – no direct evidence that 1332 actually worked against the virus in humans. So far, the drug had only been given to healthy volunteers.

It was time to test it in patients with Covid-19.

Initially, the team considered a simple study: would 1332, administered with ritonavir, reduce the viral load in patients infected with SARS-CoV-2? But that wouldn't answer the really important question: would this drug keep people with Covid-19 out of hospital and prevent them from dying? Given the urgency of the situation, Pfizer decided to proceed straight to a much bigger study comparing medical outcomes for a group of patients taking 1332/ritonavir and a group taking a placebo.

Physician James Rusnak took charge of the Phase II/III trials. Working alongside him was Jennifer Hammond. Having been so closely involved in the start of the programme, she had volunteered to help bring it to fruition.

* Ritonavir does not act directly against Covid-19; rather, it helps 1332 last longer and so act for longer.

† The drawback of using ritonavir is that if patients are already taking certain other medications metabolised by CYP3A4, this treatment may require careful management or may not be able to be administered. Fortunately, this affects only about five per cent of patients.

The trials were named EPIC: *Evaluation of Protease Inhibition for Covid-19.* The first pivotal study was a 'High Risk' trial (*EPIC-HR*), for patients who were at increased risk of progression to severe illness. The goal was to apply to the FDA for an Emergency Use Authorization (EUA), the quickest way to get the drug to patients if it proved successful.

'Given the tremendous toll Covid-19 was taking on communities around the world, we knew we needed a development programme with an ambitious global footprint,' says Rusnak. The programme would include nearly seven thousand participants from diverse backgrounds, across 350 sites in twenty countries.[5] The amount of planning and discussion required with regulatory authorities around the world was daunting. It took extraordinary efforts and coordination to open all those clinical trial sites and get supplies of the two drugs across borders in time.

Like other pharma companies, Pfizer worked closely with regulators throughout the pandemic, and decisions and agreements that might normally have taken weeks to secure were achieved within 24–48 hours. Innovative approaches were adopted to speed up some of the standard tests that regulators require of any medicine. One important mandatory test investigates how a drug is eliminated from the body. It is vital to be able to show that any foreign substance introduced into a human is eventually excreted. Normally, this test, known as a *mass balance study*, is done by attaching radioactive tags to samples of the drug given to volunteers, and then tracking those tags as they are excreted in urine or faeces. It can take a year to complete because of the complexity of the radioactive tagging process.

Amit Kalgutkar and the drug metabolism team took an alternative proposal to the FDA that would save a lot of time. As we've seen, 1332 includes an unusual molecular group called trifluoroacetamide. By good fortune, this group can be tracked, like radioactive tags, using a technique known as fluorine-19 nuclear magnetic resonance spectroscopy (NMR). Would the FDA accept a mass balance study based on fluorine-19 NMR rather than radioactive tags? Though such a test had been successfully carried out by other labs, the data had never been used in regulatory filings in place of a radioactive tag study. The FDA had questions about how the study would be calibrated but, once these were addressed, they quickly agreed. Other regulators around the world followed their example, and this neat innovation avoided months of delay.

Meanwhile, more than five hundred people worked around the clock to make enough 1332 for the clinical trials. New suppliers were contracted to manufacture large quantities of the bicyclic proline analogue and other critical ingredients. It took nine months to complete the synthesis, from commodity chemicals to clinical tablets. Still with no evidence of efficacy in humans, Pfizer had by now committed over a billion dollars to the Covid-19 antiviral programme, in addition to its huge investment in the vaccine. It was a very bold play on an uncertain prospect.

The EPIC-HR trial began in July 2021. By then, many people in the developed world had been vaccinated against Covid-19, but there were still millions who could not – or would not – take a vaccine, including many with risk factors that put them in danger of severe illness and death. Lockdowns and travel restrictions remained in force. The need for a safe and effective antiviral was immense.

The interim analysis results from EPIC-HR were released on 5 November 2021. '6:45 a.m. Eastern Time,' adds Dafydd Owen. 'They told us we had a highly efficacious molecule.'

'I remember turning on CNN early that morning, knowing the results were about to be announced publicly, and counting down the minutes before I could share the remarkable findings with the team,' Jennifer Hammond recalls.

At the time of the interim analysis, ten patients enrolled in the EPIC-HR study had died of Covid-19, all of them in the placebo arm. Forty-one were hospitalised: thirty-five in the placebo arm, and six who received 1332/ritonavir. The difference in outcomes was so tragically stark that the independent committee evaluating the results called a halt to the trial on grounds of 'overwhelming efficacy': there was now clear evidence that 1332/ritonavir worked, and therefore to continue to give sick, high-risk patients the placebo rather than the combined drug was no longer ethical.

The final results from the clinical trials showed that 1332/ritonavir reduced the likelihood of high-risk patients with Covid-19 being hospitalised or dying by eighty-six per cent, when administered within five days of symptom onset. With more than five million people dead worldwide from Covid-19 by then, this news was very welcome.[6]

1332 was given a name: *nirmatrelvir.*

The FDA granted Emergency Use Authorization (EUA) for nirmatrelvir/ritonavir on 22 December 2021, seventeen months after nirmatrelvir was first synthesised and twenty-one months after Anderson and Allerton fired the starting gun on the search for a Covid-19 antiviral. It is one of the fastest successes in the history of drug development. More than two thousand people worked night and day to make it happen. Shipments began immediately, and the team laboured through the holidays to get doses to patients in need. A little over six months later, the new medicine was administered to the President of the United States.

Full, non-emergency-use FDA approval for high-risk adults came in May 2023, by which time more than fourteen million people worldwide had already received a prescription.[7] It is probably the first medicine ever given to so many people prior to official full approval. Based upon January 2023 data,* an FDA Advisory Committee estimated that more than 1,500 deaths and 13,000 hospitalisations could be avoided with nirmatrelvir/ritonavir *each week* in the United States alone. I feel very privileged to have played a part in its final approval.

For Art Bergman, nirmatrelvir allowed the world to return to something resembling normality. The success of the drug has been

* Showing that each week in the US there were still 4,000 Covid-19-related deaths and 35,000 Covid-19 hospitalisations.

a source of great pride to him. 'It's hard to fathom that I've been involved in a project that has had so much impact.'

Dafydd Owen has to remind himself how remarkable their achievement is. '[For an individual chemist] the odds of discovering a drug are basically nil,' he says, reflecting the rarity of success in our profession. But on this one occasion, when it really mattered for global health, Owen and his team beat those odds.

The odds were improved, of course, by years of innovation across scientific disciplines and healthcare – the rapid sequencing of the virus genome, the analysis of the structure of the main protease, and the work already done on the SARS protease inhibitor in 2003. That brings us to perhaps the most important innovation lesson from this story. A lot of what we do in pharmaceutical research seems to go nowhere and can look like failure. All that work on a SARS antiviral in 2003 must have felt like a waste of time when the disease disappeared. Yet without it, we might not have arrived at nirmatrelvir so quickly. Everything we learn and discover, especially in a well-connected ecosystem with good institutional memory and data sharing, has the potential to be important at some later date. Sandeep Menon puts it best:

> You're hitting a stone twenty times; on the twenty-first time, it breaks. That doesn't mean the first twenty times were wasted. I've been in the industry eighteen years now – all those eighteen years have helped me in different ways to prepare for this moment.

Looking back on the Covid-19 pandemic, I sometimes feel disheartened at how little humanity has learned since the last great pandemic struck just over a century ago. For example, even back then, the value of masks in curbing the spread of a respiratory disease was already known. A paper published in the *Journal of the American Medical Association* in 1918 is entitled, 'The Protective Qualities of the Gauze Face Mask.'[8] By following sensible public health measures, we could have prevented at least some of the officially reported seven million deaths (as of July 2023)[9] and estimated nearly fifteen million excess deaths (as of 2021)[10] globally from Covid-19.

On the other hand, one hundred years of progress in virology, chemistry and medicine, compounded by advances in international scientific collaboration, have saved untold numbers of lives this time around. Between 50 and 100 million people died from the 1918–20 influenza pandemic, at a time when the world's population was just under two billion.[11] They had no vaccine, no antibiotics to treat secondary bacterial infections, no targeted therapeutics or ventilators. They didn't even identify the causal agent as a virus until the 1930s. We have done a lot better.

The people who made the Covid-19 vaccines and therapeutics are heroes. So are the thousands of researchers who went before them and laid the groundwork for their breakthroughs, the incredible network of investigators who oversaw the clinical trials, and the thousands of volunteers who took part in the studies. A century of slow, incremental and seemingly meandering innovation by dedicated scientists all over the world, driven by curiosity and the thirst for knowledge and solutions, brought us to the point where we could respond swiftly and successfully to the greatest pharmaceutical challenge of our lifetimes. That is what real innovation looks like: we build on the painstaking efforts of countless others and, through that accumulated knowledge, sprinkled with a dose of serendipity, we add a little something of our own to the collective endeavour.

CHAPTER 7

The Dawn of Gene Editing

Conditions: Sickle Cell Disease and Beta Thalassaemia
Molecule: Foetal Haemoglobin
Gene: BCL11A
Innovation: CTX001
Companies: CRISPR Therapeutics and Vertex

THIS IS A DETECTIVE STORY – the hunt for a tiny scrap of DNA that holds the key to relieving human suffering on a global scale. The hunt began before DNA's structure was described and ended six decades later after hundreds of scientists had pursued thousands of possible leads. The prize, that tiny scrap of DNA, is not even a gene: it does not encode a protein; it doesn't make anything. For most of us, it is completely unimportant. But this short sequence of code can drastically alter the lives of millions of people – those living with sickle cell disease or beta thalassaemia. And now that we've found that tiny scrap of DNA, the race is on to eliminate it.

People with sickle cell disease suffer terrible pain crises. When symptoms are severe, they will scream in agony. Some must be hospitalised for pain control. Such episodes can occur multiple times a year. Parents forced to witness a full-blown sickle cell pain crisis in a young child find the experience unbearable.

Sickle cell disease is an inherited, genetic condition that was first described scientifically in 1910. Walter Clement Noel was born on the Caribbean island of Grenada. Of African descent, his family were well-off landowners. Noel received a good education and was able to travel to Chicago to study dentistry.[1] While in the United States, he suffered a series of pain crises and consulted a physician named James B. Herrick. Although Herrick, a cardiologist, usually gets the credit for the discovery, it was his resident, Ernest Irons, who examined Noel. He took a blood sample and under the microscope saw red blood cells that had become distorted. Healthy 'red cells', as we tend to abbreviate them, are soft and round, resembling a doughnut without a hole. Noel's red cells were elongated and hard-edged, like a crescent with pointed ends – 'having the shape of a sickle,' as Irons wrote.[2] Reading this later, Herrick postulated a new disease and published the finding, securing his place in medical history.

The primary job of a red cell is to carry oxygen around the body. Oxygen molecules bind to haemoglobin in red cells in the lungs, and then are released in various organs as the cells circulate. Sickle cells are less efficient at carrying oxygen than normal red cells. That in itself is not a terrible problem, but sickle cells also have a tendency to clog blood vessels, blocking the flow of blood and the provision of oxygen to tissues. This leads to excruciating pain, and it can start happening in infants from the age of about five months.

Beyond the pain, the oxygen deprivation can lead to organ and bone damage, strokes, pulmonary crises and premature death. Doctors may have to remove the abnormally enlarged spleen of patients whose blood flow to the organ has become clogged. Hip damage is common, as the blood vessels here are small and easily blocked, leading to bone death. Patients have low red cell counts (anaemia), exacerbating the oxygen deprivation.

In 1910, Walter Clement Noel was treated with warmth, fluids and some pain control. 'Even today, in 2022,' observes sickle cell physician Bill Hobbs, 'that's still the way we treat an acute pain event in sickle cell disease. Over a hundred years later, we haven't really got much better at treating an acute pain episode.'

Millions of people suffer from sickle cell disease. Most are African, or of African descent, though it also affects some people with southern European, Asian and Middle Eastern heritages. Just in Nigeria, around 150,000 babies are born each year with the disease.[3] Sickle cell disease is a recessive genetic condition, meaning that only people who inherit two copies of the sickle cell version of the *HBB* gene (which encodes beta haemoglobin) have the disease. People with just one copy don't have the same symptoms, although their haemoglobin is slightly different, and they are said to have *sickle cell trait*.

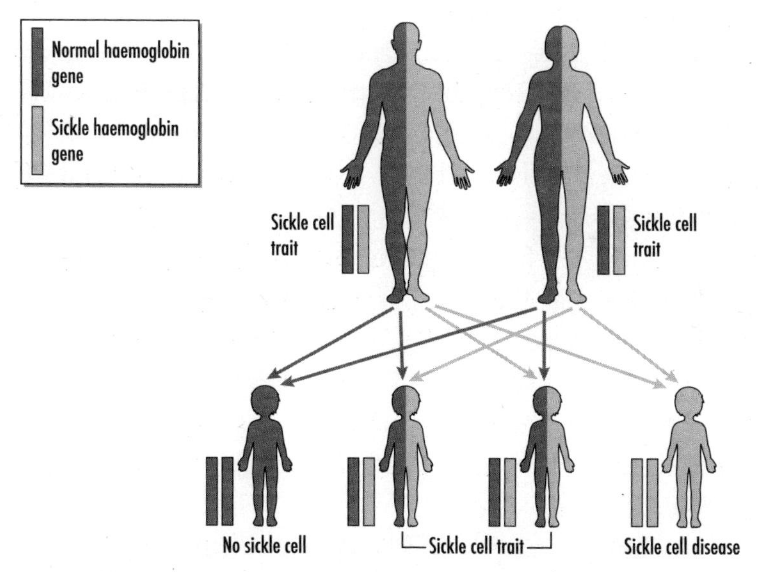

Figure 18. Sickle cell disease is an autosomal recessive disease.

It is unusual and strange that such a debilitating condition should spread so widely in a population. Evolution doesn't usually do that: genes that result in premature death tend to get weeded out over time by natural selection, especially if those affected do not live to be old enough to have children. So why was the gene for such a damaging disease so widespread in Africa?

* * *

Anthony Allison grew up on a chrysanthemum farm overlooking Kenya's Great Rift Valley during the height of the British colonial period. At the time, paleoanthropologist Louis Leakey was making remarkable discoveries in the region, unearthing what seemed to be our humanoid ancestors. After meeting the great fossil hunter, Allison developed a passion for evolution and anthropology.[4]

In 1949, Allison joined a University of Oxford research expedition to Kenya. Most of the scientists were studying local plants or insects, but Allison had developed a fascination for human genetics as a way to understand – as Leakey was doing with bones – the origins and evolution of human populations. He travelled all over Kenya, collecting blood samples from different tribes, and mapping the incidence of various traits.

One of them was sickle cell trait.

Having enjoyed a childhood in the salubrious Kenyan highlands, Allison now saw a different side of the country. In hospitals in Mombasa and Kisumu, the children's wards were full of boys and girls in terrible pain, victims of sickle cell disease.[5] This grim condition had been far less prevalent where Allison grew up. His blood samples told the same story: 'I found that among tribes living close to the coast of Kenya or to Lake Victoria, the [sickle cell trait] frequencies exceeded 20%, whereas among several tribes living in the Kenya highlands or in arid country, the frequencies were less than 1%,' he wrote. 'These differences cut across linguistic and cultural boundaries.'[6]

In other words, something seemed to be driving the selection of the sickle cell trait in humans living at lower altitudes and in areas of higher rainfall.

'I formulated an exciting hypothesis,' wrote Allison. People with sickle cell trait would, he suggested, 'have a selective advantage, because they are relatively resistant to malaria.'

It was a simple, brilliant idea. Sickle cell disease may be bad, but malaria is worse. Moreover, most people carry just one copy of the sickle cell version of the *HBB* gene – they have sickle cell trait – and so don't suffer the debilitating symptoms of the disease. The mosquitoes that carry the malarial parasite are most common at low altitude, and they need standing water to breed. Malaria is more prevalent along the humid Kenyan coast and around Lake Victoria, where Allison had found high frequencies of sickle cell trait.

The selective mechanism wasn't hard to envisage. *Plasmodium falciparum*, the protozoan (single-celled microscopic organism) that causes the most dangerous form of malaria, lives for part of its life cycle in human red cells. If those cells are distorted, or the haemoglobin is abnormal, perhaps the malarial parasite would struggle to survive. That might offer enough of a selective advantage, in a community where babies routinely died of malaria, to cause the sickle cell trait to spread.

In fact, Allison wasn't the only scientist considering the power of malaria as a selective agent that year. Another inherited blood disease, *thalassaemia*, results in anaemia and inadequate oxygen provision to tissues. This can also cause terrible pain, destruction of bones, facial deformation and death. Thalassaemia was historically common around the Mediterranean, so the same question arose: why would the gene for this nasty disease spread widely in southern Europe? The evolutionary geneticist J.B.S. Haldane offered his answer* in a 1949 paper: referring to the undersized red cells found in thalassaemia patients, he wrote, 'It is at least conceivable that they are also more resistant to attacks by the sporozoa which cause malaria, a disease prevalent in Italy, Sicily and Greece, where the gene is frequent.'[7]

Malaria was the problem. Thalassaemia was nature's response in the Mediterranean. Sickle cell was the response in Africa.

Allison was able to gather experimental evidence to back up his hypothesis in 1953. A pharmaceutical company was testing new antimalarial drugs at a laboratory in Kenya by first infecting volunteers with *Plasmodium falciparum*. This is known as a challenge study (an alternative, proven antimalarial was available to treat

* Haldane may have got the idea from Italian geneticist Giuseppe Montalenti.

anyone who became ill). The deliberate infections gave Allison an opportunity to measure malarial resistance in people with sickle cell trait. Those volunteers who had the trait showed some resistance, with lower parasite counts in their blood.

Concerned that this finding might be compromised by the effects of acquired immunity in adults, Allison then took blood samples from young children in Uganda; before the age of four, children would be unlikely to have built up much immunity from previous infections. He found that those children with sickle cell trait had lower malarial parasite counts. Allison concluded that children with sickle cell trait 'are more likely to survive through early childhood in a highly malarious environment.'[8] Other researchers later confirmed Allison's findings in a range of African populations.* When maps of sickle cell incidence and malaria prevalence were drawn up and compared, they were closely aligned.

While Allison was studying the epidemiology of sickle cell disease, other scientists were elucidating its molecular basis. Haemoglobin is made up of four polypeptide subunits – two alpha (α) subunits and two beta (β) subunits. Each of the four subunits has an iron-containing *heme* molecule which binds oxygen. In 1949, Linus Pauling, one of the founders of molecular biology, discovered that the haemoglobin molecules in sickle cells were elongated and distorted.[9] Further work showed that the beta-globin subunits had a different structural form in sickle cells. Pauling called sickle cell disease the first 'molecular disease', marking the moment when a defective protein was first associated with a human disease.

In 1956, Vernon Ingram, a postdoctoral fellow in Pauling's lab, found that this sickling form was caused by a single amino acid substitution in beta-globin.[10] This in turn was caused by a

* Although the actual mechanism of malaria resistance is still being debated. See: https://doi.org/10.1371/journal.pmed.0020128.

single-letter mutation in the gene encoding beta-globin, *HBB*. In medical school in the 1990s, we all learned about this classic example of a single gene mutation causing a devastating disease. People with just one copy of the mutated gene – carriers with sickle cell trait – produce both normal and abnormal haemoglobin. The former keeps them healthy; the latter offers protection against malaria.

Unfortunately, this growing body of knowledge offered little comfort to people suffering from the disease's agonising symptoms. Genetic diseases were not considered fixable until recently. As we saw in Chapter 1, physicians presumed that anything written in the genes was a permanent endowment. Once someone was born with two copies of the sickle cell version of the *HBB* gene, there was nothing that could be done for them other than pain relief and blood transfusion.

That presumption was blown out of the water, at least conceptually, by the work of a paediatric haematologist in New York. In 1948, Janet Watson noticed something curious about children born with sickle cell disease: they did not suffer its symptoms straight away. Newborns were spared the pain crises that would afflict them for the rest of their lives. Analysis of blood samples revealed that most of their red cells had not yet sickled. Even though they were genetically programmed to have sickle cell disease, their cells – at this very early stage – were mostly normal.

Watson came up with a hypothesis to explain the discrepancy.[11] She knew that different types of haemoglobin are formed at different stages of human development: embryos have *embryonic haemoglobin*; foetuses have *foetal haemoglobin*; and after birth we produce *adult haemoglobin*.

Could it be, she speculated, that foetal haemoglobin – still flowing in the veins of newborns – was somehow immune to sickling?

Innovation journeys often start with a simple observation that sparks an idea. For the treatment of sickle cell disease, that idea was foetal haemoglobin.

* * *

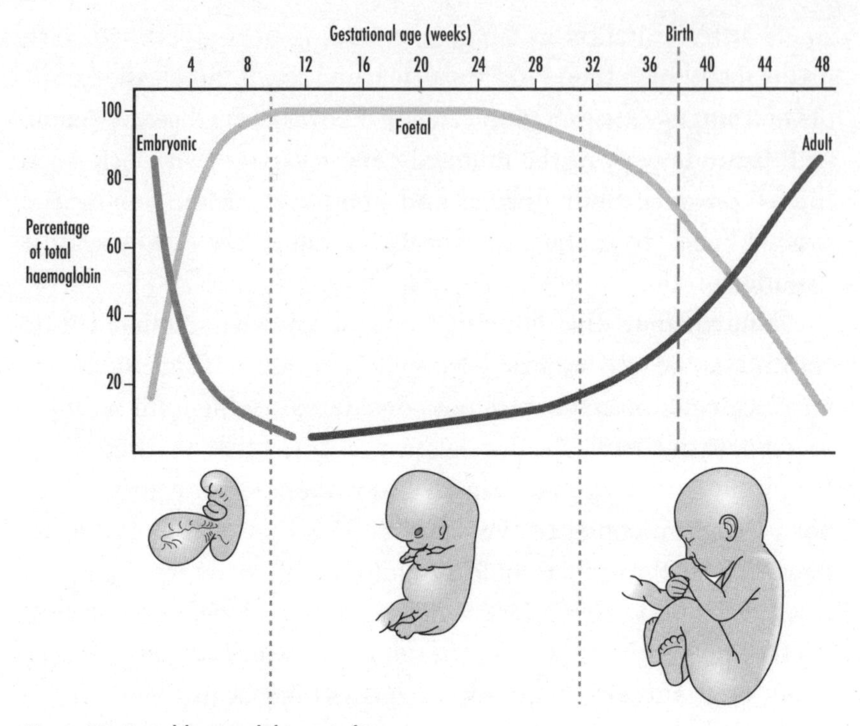

Figure 19. Foetal haemoglobin switching.

George Stamatoyannopoulos was possessed of such a challenging name that stumped colleagues usually referred to him as plain 'Stam'. 'You need a bit of a run-up at that name,' quips Doug Higgs, one of the foremost experts on globin gene regulation. Luckily Stamatoyannopoulos became such a big figure in the world of inherited blood diseases that a single syllable sufficed to identify him.

'He was a titan of science,' says Fyodor Urnov, a professor at Berkeley with a focus on genome editing. 'In the classic, sacred tradition of great European Enlightenment Age wisdom.'

Stamatoyannopoulos's success and stature were all the more remarkable for his bleak start in life: born in Athens in 1934, he grew up under Nazi occupation, then endured famine and a civil war that killed 100,000 Greeks and displaced nearly one million. Through this chaos, Stam persevered at his studies, entering medical school at the age of seventeen.[12] He graduated at the top of his class in 1958.[13]

A geneticist at heart, Stam was drawn to inherited blood disorders, and he focused his initial research efforts on thalassaemia, a common disease in Greece. During his military service, he would walk to remote villages in the Greek mountains and islands, where he would construct family histories and take blood samples from people with thalassaemia.[14] We can imagine the young scientist wandering along dusty tracks scented with pine and wild oregano, the baking air alive with the screech of cicadas, in search of genetic answers to an age-old affliction.

'He would convince the locals during the religious holidays when they were not allowed to eat meat that he needed to sacrifice some sheep to do blood analysis,' says Urnov. 'And he just ate the sheep.'

With Janet Watson's 1948 observation in mind, Stam began to look into foetal haemoglobin. Was she right to think that foetal haemoglobin was somehow immune to sickle cell disease, he wondered, and if so could the same be true for thalassaemia?

Sometimes, the critical question in science is, *How can we know?* Was there any possible way – in 1960s Greece – to show that foetal haemoglobin had special powers? In biology, the answer to difficult questions is often revealed through unusual cases, and in medical research that means rare diseases. Often, the study of rare diseases has led to insights into more common maladies.

In 1961, a Greek family had been found to have a condition called hereditary persistence of foetal haemoglobin (HPFH).[15] Most of us have only tiny amounts of foetal haemoglobin in our blood, but people with HPFH keep making foetal haemoglobin in substantial quantities long after birth. It does them no harm – foetal haemoglobin works just as well as adult haemoglobin – but it is unusual.

When he discovered that some members of the HPFH family also had thalassaemia, Stam realised that this rare condition opened an exceptional window on to the interaction between foetal haemoglobin and beta-globin diseases like sickle cell disease and thalassaemia. With physician and haematologist Phaedon

Fessas, he embarked on a study of those few people who have both HPFH and thalassaemia.

Levels of foetal haemoglobin vary considerably in people with HPFH. The two researchers were able to show that, for people with thalassaemia, the more foetal haemoglobin they produced the less severe were their thalassaemia symptoms.[16]

It was a crucial breakthrough. Janet Watson had suggested that foetal haemoglobin might somehow be resistant to the debilitating distortion of sickle cell disease. Now, Stam and Fessas had shown that people with high levels of foetal haemoglobin were protected against the worst symptoms of thalassaemia. Foetal haemoglobin was starting to look like a natural miracle drug for people who had inherited one of these awful diseases.

Most of us stop making foetal haemoglobin when we are born: the body switches production from foetal haemoglobin to adult haemoglobin. Generally, that doesn't matter – we don't need foetal haemoglobin if we have normal adult haemoglobin. But for those with compromised adult haemoglobin, it's as if their miracle drug is cut off soon after birth.

It wasn't long before Stam made the pivotal conceptual leap: might it be possible to reverse that switch and turn the production of foetal haemoglobin back on? And if so, would that effectively *cure* people with thalassaemia and sickle cell disease?*

What's the difference between foetal haemoglobin and adult haemoglobin? Functionally, very little. Adult haemoglobin releases oxygen a little more easily than foetal haemoglobin (thus enabling a foetus to acquire oxygen from maternal blood), but otherwise the two do much the same thing. Structurally, there is one important difference. In adult haemoglobin, there are two alpha subunits and two *beta* subunits. In foetal haemoglobin, there are two alpha subunits and two *gamma* subunits.

* Other scientists were thinking along similar lines in the 1960s, including two renowned haematologists, David Weatherall and David Nathan.

Beta thalassaemia – the type common in Greece – occurs when there is an imbalance between alpha (too many) and beta (too few) subunits, leading ultimately to the loss of red cells. Sickle cell disease occurs when a mutation in the beta-globin gene results in the synthesis of abnormal beta subunits. In both diseases, it is the beta subunit encoded by the beta-globin gene that is the problem. So you can see why foetal haemoglobin, which does not have a subunit encoded by the beta-globin gene, is unaffected by either disease.

The big question that George Stamatoyannopoulos and others continued to wrestle with was: *how does the body switch production from foetal haemoglobin to adult haemoglobin at birth, and how might that switch be reversed?*

Stam had moved to the United States in 1964 to take up a position at the University of Washington. Over the subsequent years, he built up a network of like-minded scientists interested in the switching problem, and in 1978 he convened, with NIH haematologist Arthur Nienhuis, a meeting of researchers from a range of disciplines to investigate foetal haemoglobin switching. Every year, at least 300,000 people were born with severe forms of sickle cell disease or thalassaemia; the two men wanted to focus minds on this possible solution. The conference took place at the leafy Battelle Memorial Institute in Seattle. It would be the first in a series of biennial meetings, funded by the NIH, that continues to this day.

That first meeting brought together a small, rather exclusive group of around fifty scientists. One was Tom Maniatis, a pioneer of gene cloning. At the time, his lab was cloning the haemoglobin genes, and he presented the first announcement of this breakthrough at the Seattle conference. It was a landmark moment for science, and it helped put the switching meetings on the map.

Among the other participants was a young physician-scientist, Stuart Orkin. He'd been working on the molecular biology of haemoglobin, and when Stam's invitation arrived he was attempting to map the haemoglobin genes – that is, determine their physical location on one or more of our chromosomes.

Orkin listened attentively to the presentations from David Weatherall, George Stamatoyannopoulos, Phaedon Fessas, Vernon Ingram, David Nathan and all the other greats in the field. He himself presented his own mapping work. He drank up all the latest developments in cellular and molecular biology that might be relevant, and saw the huge opportunity if only the switching mechanism could be cracked.

By the end of the conference, he was hooked.

'Once he gets his teeth into something, he goes through it like a hot knife through butter,' Doug Higgs says of Orkin. 'He's enormously effective.'

Stuart Orkin grew up in Manhattan, where he devoted much of his time to baseball and the New York Yankees. 'I was not the most obedient kid. I was bored in a lot of classes. I didn't get the best grade in citizenship.' Now in his seventies, Orkin enjoys classical music, family gatherings and travel. His euphemism for dying is 'going permanently to Hawaii'.

Orkin's father was a surgeon, but it wasn't immediately obvious that he too would pursue a medical career. In high school he found he had an aptitude for physics, and that took him to MIT. But at college he found himself surrounded by classmates who, as he puts it, 'could intuit physics where I needed to use a pencil and paper'. He felt outclassed by his peers and put an end to his physicist dream.

Instead, he tried a molecular biology course and was lucky enough to be taught by Salvador Luria, who would later win a Nobel Prize for his work on the genetics of viruses. Graduating in 1967, unsure what to do next, he eventually opted to follow his father into medicine.

The Vietnam War was raging while Orkin was at Harvard Medical School, and the threat of the draft loomed: medically trained men were in high demand on the battlefield. But for some, there was an alternative: the 'Yellow Berets', as these physicians were sometimes unkindly known, could instead take a position in the US Public Health Service. A privileged subset could apply to join the clinical associate programme at the NIH. My postdoctoral fellowship mentor, Harold Varmus, took this route, and for him and several others it laid the foundations of a Nobel-Prize-winning career. Anthony Fauci, world-famous for his leadership of the National Institute of Allergy and Infectious Diseases, was another clinical associate. In retrospect, the Vietnam-avoiding programme was a wonderful incubator for some of our very best physician-scientists.

Stuart Orkin was one of them. 'That was the heyday of NIH medical training,' he reminisces. In those days, medical schools didn't have the research infrastructure they do now, so the NIH played an outsized role in nurturing great medical science. Orkin applied to join the lab of Phil Leder, the remarkable molecular geneticist who, with Marshall Nirenberg, helped decipher the genetic code.

Leder's group was working on the globin system and immunoglobulins. As a medical student, Orkin had already developed an interest in blood, and he enthusiastically joined in the globin research. 'I liked blood. It was an accessible system with a clear relationship to disease. There were a lot of disorders of blood that were clearly genetic, and that at least in principle could be worked out.'

After the NIH, Orkin became a fellow and faculty member at Boston Children's Hospital, where he was strongly influenced by the Director of Haematology, David Nathan.

Nathan was 'Mr Haematology', in Orkin's judgement. 'Nathan had the vision of bringing molecular biology to the problem of thalassaemia.' It was Nathan who first proposed using hydroxyurea, a chemotherapy drug, to boost foetal haemoglobin; hydroxyurea is still one of the few approved treatments for sickle cell disease.

On the paediatric haematology ward, Orkin witnessed first-hand what the disease meant for young children. The sheer agony they endured struck him profoundly, particularly as there was so little he could do for them. In conversations with Nathan, he began to understand the critical role and potential promise of foetal haemoglobin for people with thalassaemia and sickle cell disease.

So began a decades-long fascination and research effort.

The problem was that no one had a clue what mechanism caused the switch from foetal to adult haemoglobin. Was there a gene that switched off foetal haemoglobin production, or one that switched on adult haemoglobin production, or several genes that interacted? Or was it something else? Could something in the environment drive the switch?

'We didn't know what we were looking for,' says Orkin.[17]

Following the inaugural switching meeting, Orkin drew on the new art of cloning to work out the genetic basis of beta thalassaemia. With physician-geneticist Haig Kazazian, he examined and described the different genetic mutations that caused beta thalassaemia in a range of ethnicities. Together, they isolated mutant beta-globin genes and then Orkin cloned and sequenced them.

'What fun this was!' Kazazian later recalled. 'Stu would call almost daily with new information! He stayed at our home as we finished the first paper in late 1981. Over the next five years, we wrote twenty-five other papers covering beta [thalassaemia] mutations in other ethnic groups.'[18]

Ultimately, they published an almost complete catalogue of genetic mutations causing beta thalassaemia. 'I think it was probably the first disease for which molecular biology had given the full description of the mutation pathology,' says Orkin. Their findings have been profoundly important for prenatal diagnosis: we can now test human embryos for the presence of any of these mutations.

It was a huge achievement that established his reputation in haematological circles, but Stuart Orkin sets himself high expectations: 'It was nice, but I was a little disappointed because we didn't learn anything about red cells.' All that work had taken him no closer to the goal of understanding haemoglobin switching. So he changed gears and started looking at transcription factors.

A transcription factor is a protein that regulates how DNA is transcribed into RNA – in other words, it controls how and when genes are expressed. Transcription factors operate by binding to a particular (non-gene-encoding) DNA sequence, something that will become important later in this story. Orkin wanted to understand which transcription factors were involved in the development of red cells.

In 1989, he discovered the first red cell transcription factor, now called GATA-1.* This is the 'master controller' for red cell genes: it regulates the maturation of red cells and platelets, among many other functions. At first, Orkin believed it was involved in foetal haemoglobin regulation. But further research showed that GATA-1 has no stage specificity – it's expressed at embryonic stage, foetal stage and adult stage – so it couldn't explain a stage-specific switch like the change in haemoglobin production after birth.

Again, although the discovery of GATA-1 ranks as a tremendous scientific feat, Stuart Orkin was disappointed in its failure to advance his foetal haemoglobin quest. He would make no further progress on that front for nearly two decades.

'For the entirety of the nineties and the early 2000s, the field was kind of lost,' says Orkin. Research into foetal haemoglobin switching continued in his lab and others, but no one could make headway. Some scientists tried removing the pituitary gland in animal models, others went for the adrenal glands, hypothesising that some circulating substance released by these organs might be controlling the switch. Various drugs were tried. Nothing restored the flow of foetal haemoglobin.

'There was a huge amount of work going on, and we didn't really get very far,' says Doug Higgs.

People grew discouraged, and although George Stamatoyannopoulos resolutely kept the biennial switching meetings going, they tended to focus more on adjacent areas of research

* Another researcher, Gary Felsenfeld, made the same discovery independently, around the same time.

such as gene therapy.

Orkin's lab discovered a second transcription factor, but that was no more relevant than GATA-1. Another group discovered a transcription factor that was important for beta-globin gene expression, and Orkin investigated that for a while, finding only a peripheral connection with haemoglobin switching.

'Things were kind of stuck,' he concluded. Nothing they looked into explained how the body switched production of foetal haemoglobin off, and there were only so many leads to pursue. 'We ran out of things to do.'

The Orkin lab had plenty of other projects under way, and these steadily displaced the haemoglobin work. To Orkin's profound regret, haemoglobin switching disappeared from the agenda. 'I didn't forget about it, but the problem is trainees in the lab don't want to work on something that isn't going anywhere.'

Paradoxically, it would be one of those trainees who finally brought Orkin's switching programme back from oblivion.

Vijay Sankaran grew up outside Boston. At the University of Pennsylvania, he studied mostly chemistry and physics. 'I was probably the wrong kind of person to think about a career in biology,' he laughs. Nevertheless, he joined a biophysics lab and found that he loved thinking about biology at the molecular level. His passion for science is clear: he is constantly in motion, even while sitting down. His arms gesture wildly, he rocks back and forth. His infectious good humour animates his whole frame.

Enrolling as an MD PhD student at Harvard Medical School in 2003, he met Stuart Orkin at the nadir of haemoglobin switching research. The decades-long effort had finally run out of road. Still, Sankaran was looking for a mentor, and Orkin fit the bill. During a training rotation in the Orkin lab, Sankaran worked on embryonic stem cells (cells that have the potential to develop into any kind of cell). A small accident blighted his early experience there: while bowling, he slipped and fractured his wrist. 'So, as you can imagine, being in the lab, that was a little bit limiting, having a cast on my wrist.'

Nevertheless, Sankaran felt at home in the Orkin lab and decided to join it full-time for his PhD. But he wasn't drawn to the stem cell work that constituted the lab's main focus. As part of his medical training, Sankaran had spent time at a paediatric haematology clinic, where some of the patients had sickle cell disease. 'One of them was a little older than I was. He'd endured many, many pain crises,' recalls Sankaran. 'It was one of my first experiences where I was just left in the room with a patient and I had an hour to talk to them, to get to know them.'

He was aware that Orkin had devoted years of study to thalassaemia and sickle cell disease, and that he had largely given up. 'The lab was an exciting place . . . but no one was working on this.' His curiosity piqued, Sankaran started reading through the old papers on foetal haemoglobin. He found a long piece on haemoglobin switching written by George Stamatoyannopoulos, which laid out in detail the age-old challenge. Inspired, he asked for a meeting with David Nathan. Although then in his seventies, Nathan had lost none of his determination to tackle blood diseases. 'He was just getting so excited, he was banging on the desk.' During their two-hour conversation, Nathan's passion took root in Sankaran.

So in late 2004, Sankaran went to Orkin and said, 'Look, this is really what I want to do.'

'I warned him that most people who chose to work on haemoglobin switching ended up losing their funding,' Orkin laughs.

But for Orkin, it was like a sign from above. This young researcher, this mere student, was pushing him to get back in the saddle. 'Before Vijay came along, Stu was out of it – he'd given up,' says Doug Higgs. 'He thought it was a hopeless task.'

Orkin and Sankaran sat down to work out how to come at the problem from a fresh angle. They decided to investigate haemoglobin switching using transgenic mouse models – mice genetically modified to develop human beta-globin. Sankaran followed this line of enquiry for the next couple of years, but nothing worked. It turned out there was a critical difference between mouse and human gene regulation that undermined the entire research effort.

Undaunted, Sankaran tried again. His next approach focused on a rare form of childhood leukaemia which is associated with elevated foetal haemoglobin levels. The cancer is caused by an oncogene called RAS, and Sankaran investigated whether the same gene might also control haemoglobin switching. That too led nowhere.

'Those first few years were really frustrating,' says Sankaran. 'I was like, "OK, I'm going to finish my PhD and not have a single paper." '

They were lonely years, too – he was the only person in the whole lab working on haemoglobin. All of Orkin's other associates were busy with stem cell research.

'I went through two years with nothing,' he says. 'Stu was very supportive, we would meet regularly, but there wasn't much to report.'

Then, a new possible lead. Back in the 1990s, a researcher at Johns Hopkins had found what seemed to be a link between higher foetal haemoglobin levels and the X chromosome. If the link was substantiated, it could really narrow down the search for a possible gene controlling haemoglobin switching.

To test the putative link, Sankaran applied for access to a collection of around a thousand DNA samples from patients with sickle cell disease that had been stored on ice since the 1980s by the NIH. 'At the time, I don't think anyone was requesting it. There was this resource – no one cared,' laughs Sankaran. 'It had been sitting around in an NIH freezer for decades.'

Sankaran studied known genetic markers on the X chromosome to see if any were especially common in the sickle cell DNA samples. There was no 'signal' at all in his results. Sickle cell disease, it seemed, was not associated with any particular variation on the X chromosome. 'This is not working,' Sankaran told himself, and he put the DNA samples aside. The X chromosome lead was dead. It was one more frustration, one more failure.

* * *

Left-field ideas are, by definition, unusual, but in innovation they can be transformational. Sometimes the best brains in science will work away at a problem, pursuing every possible lead in their universe, until nothing is left to try and their immense effort grinds to a halt. Then something happens in a completely different universe – left field – and a whole new line of enquiry opens up.

Orkin and Sankaran had done everything they could to investigate the *molecular biology* of foetal haemoglobin switching. The answer that had eluded them would come in the end from an entirely different academic discipline: *population genetics*.

Picture two distinct scientific beasts. One is a lab researcher like Orkin or Sankaran, hunched over a microscope or a gel electrophoresis system. The other is a wandering surveyor, clipboard in her hand, sun cream and mosquito repellent in her bag, measuring strangers' heights and asking about their disease history. We've already met this species of scientist in George Stamatoyannopoulos and Anthony Allison. Their tools – questionnaires, charm, diplomacy and statistics – are about to unlock the genetic secret that Stuart Orkin had pursued for so many years.

Population geneticists look for shared traits (specific characteristics of individuals) in people and then search for small variations in their DNA that correlate with those traits. In this way, they hope to connect a particular gene or genes with a trait.

In the early years of the discipline, researchers would hypothesise that a particular gene had something to do with a trait like diabetes or green eyes, and would then scan that gene in people with that trait to see if they shared a genetic sequence that others did not. This *candidate-gene association* approach was challenging, often because researchers were not good at guessing which genes might be relevant. There were still large areas of the genome that science knew nothing about.

But by the early 2000s, gene sequencing had advanced so fast that it was now possible to search the entire genome – all the genes in a person – for shared genetic variations. This opened the door to a groundbreaking new scientific tool: the *genome-wide association study* or GWAS. It was no longer necessary to guess which genes might influence, for example, height. Researchers could select the tallest people from a population and see if they shared unusual genetic sequences *anywhere* in their genome.

The most common genetic variations are called single-nucleotide polymorphisms (SNPs), pronounced *snips*. SNPs occur throughout our DNA, with 4–5 million SNPs in each individual's genome.* Most SNPs occur in the DNA *between* genes. To be classified as a SNP, a variant must be found in at least one per cent of the population. So, at one place in the human genome, you might have a *C* while your neighbour has a *T*. The rest of the surrounding genetic sequence is the same for both of you. What effect does this difference have? In most cases, there appears to be no effect. Some SNPs can affect our risk of developing diseases and may influence how we respond to certain drugs.[19]

Trying to understand the impact of a SNP is very hard to do in the lab, though. In most cases, we can't simply find out what a particular letter in a DNA sequence 'does'. But another way to come at the problem is to look at lots of people with the same SNP and see if they share some particular trait that others don't. It's not proof the SNP is responsible for the trait, but the correlation may be suggestive of some causal connection.

Gathering the data to form reliable correlations means talking to a very large number of people. This is where our population geneticist with the clipboard and the sun cream comes into play. A GWAS looks at a population of thousands, recording dozens of traits from each individual and sequencing their DNA. It's then up to the statisticians to figure out whether enough people in the study share both a trait and a particular SNP to suggest a probable genetic link.

One of the earliest full-scale genome-wide association studies took place in Sardinia.

* * *

* The human genome contains about three billion nucleotide (A, C, G, T) pairs.

The Italian island of Sardinia used to have a big beta thalassaemia problem. 'It was everywhere,' says Bastiano Sanna, a pharmacologist who grew up on the island. Untreated thalassaemia patients looked different, he explains. Their distorted bone structure made their cheeks droop and their upper teeth protrude. 'In Sardinia, you could not escape thalassaemia.'

This is no longer the case, thanks in part to Stuart Orkin's work on thalassaemia mutations, and in part to a brilliant local doctor and geneticist called Antonio Cao, who instituted a prenatal screening and genetic counselling programme across the entire island.

'It's been a triumph of public health administration,' says Stuart Orkin. Across Sardinia, the incidence of beta thalassaemia plummeted from one in 250 people in 1975 to just one in 4,000 twenty years later.[20] Similar mass screening programmes have been successfully conducted in other thalassaemia hotspots, such as Montreal in Canada, Marseilles in France, and the island of Cyprus.

From his island-wide screening programme, Cao accumulated a great deal of data about the physical and physiological traits of thousands of people. In this immense population dataset, he had the perfect foundation for a GWAS. He just didn't know how to go about conducting it.

David Schlessinger has been working in the fields of microbiology and genetics for more than sixty years. He ran the Human Genome Centre at Washington University, then moved to the National Institute on Aging (NIA) in 1997, 'both as an example and as a researcher'. There he set up a new genetics department, including a unit focused on statistical genetics, a novel field at the time.

Schlessinger was training a postdoc, Giuseppe Pilia, who came from Sardinia and had previously studied under Antonio Cao. 'He was an extraordinary character, very charismatic,' says Schlessinger of Pilia. 'He was quite brilliant, perhaps the most brilliant postdoctoral fellow I've ever had.' Among his other achievements, Pilia mapped part of the X chromosome and discovered the genetic basis for several diseases.[21] One day, he went to see Schlessinger and suggested Sardinia might make for a great population study of ageing and other traits.

'That seemed to me quite impractical,' Schlessinger says, 'but as we talked, it became more and more attractive.'

Sardinia is an unusual case, because most of the modern population stems from a very small *founder group* of settlers. Consequently, there is far less genetic diversity – which could contribute to statistical *noise* – to muddy population genetics studies. For example, the beta thalassaemia that was widespread on the island was almost all caused by the same genetic mutation, whereas in other locations a multitude of mutations can cause the disease. Schlessinger gave Pilia the go-ahead, and the NIA agreed to fund the GWAS.

Previous attempts to run population studies on Sardinia had been rejected by its inhabitants. But Pilia was able to enlist Cao's help in winning over the key opinion leaders, including the local bishop and mayors. Pilia's family was well known and liked, and they encouraged friends and neighbours to come on board. 'His father was the local radiologist, his mother was a grade school teacher who had taught everyone in town,' laughs Schlessinger. The GWAS would cover diseases like diabetes, thalassaemia and multiple sclerosis, which were particularly prevalent in Sardinia, as well as traits like height and ageing-related conditions.

Pilia focused on four towns in one valley in south-eastern Sardinia, including his home town of Lanusei. The area is known as an 'island within the island', as it is surrounded on three sides by mountains and on the fourth by the sea. That made it very isolated, and the genetics of its people unusually homogeneous.

It took Pilia four years to sign up around six thousand volunteers. He collected blood samples and recorded trait data, and had their genomes analysed using *DNA chips*, microarrays of DNA spots on a postage-stamp-sized plate that allow simultaneous testing of large numbers of SNPs.[22]

Tragically, while the genome analysis was under way, Pilia developed lymphoma. He died in 2005, aged just forty-three. 'He was extraordinarily strong and vigorous, so it was shocking to everyone,' says Schlessinger.

Nevertheless, the results started to roll in. Gonçalo Abecasis, a biostatistician at the University of Michigan, wrote software to analyse Pilia's data and identify statistically valid associations between genes and traits. His programme detected possible associations for a variety of traits, but one stood out. It was something no one had expected. It concerned foetal haemoglobin.

Pilia had measured the haemoglobin levels in his blood samples, and now Abecasis used the data collected by his team to show high foetal haemoglobin levels were strongly associated with particular SNPs in a gene called *BCL11A*. This gene was known to be highly expressed in the brain, but had never previously been suspected of having anything to do with haemoglobin.

The GWAS had turned up a truly unanticipated and fascinating result.

'There had been a generation of very talented biochemists and haematologists trying to figure out what accounted for the haemoglobin switch,' says Schlessinger. 'Lots of things had been tried and *nothing* had worked.' He is smiling proudly as he delivers the GWAS punchline: 'Instead, the agnostic genetic testing pointed directly to that gene.'

'We would never in a million years have guessed it,' admitted Orkin later.[23] There was absolutely no reason in the lab to imagine that this gene, out of so many others, was connected to foetal haemoglobin switching. Only a population study – focused mostly on other traits – was able to reveal it.

Schlessinger's group conducted a further study on *BCL11A*. A subset of thalassaemia patients have a mild form of the disease, called thalassaemia intermedia. These patients tend to have higher levels of foetal haemoglobin, explaining the less severe disease. Schlessinger was able to show that all such patients within the Sardinia population had the *BCL11A* SNPs. It wasn't proof, but it was a great indication that some kind of intervention in *BCL11A* might help lessen the severity of thalassaemia in other patients.

The Sardinia group submitted a paper to the *New England Journal of Medicine* in 2007, describing the association between *BCL11A* and higher levels of foetal haemoglobin. To their surprise, it was rejected. One of the comments from the reviewers was that they should confirm the result by looking for the association in a population of sickle cell disease patients who tend to have elevated levels of foetal haemoglobin.

Dismayed, Abecasis relayed this comment to another GWAS pioneer, an endocrinologist at Boston Children's Hospital. Abecasis was pessimistic about their prospects: assembling a population of sickle cell patients for another GWAS would be no small undertaking. But the endocrinologist suggested an alternative approach. He knew of a sickle cell DNA collection, just down the corridor from his office – the very same collection Vijay Sankaran had recently pushed to the back of the freezer in the Orkin lab.

Already excited by the news that a possible switching gene had been found, Sankaran readily agreed to analyse the NIH cohort. Retrieving the DNA from the freezer, he tested the samples and found that, in sickle cell patients too, the *BCL11A* SNPs were associated with higher levels of foetal haemoglobin. 'We replicated those associations beautifully,' he says.

With two association studies – one in a 'healthy' population and one in a sickle cell cohort – both concluding that there was a strong association between *BCL11A* variations and elevated foetal haemoglobin, the combined team of David Schlessinger, Gonçalo Abecasis and Antonio Cao on the population genetics bench, and Vijay Sankaran and Stuart Orkin on the molecular biology bench, went back to the *New England Journal of Medicine* with their joint results. Still the journal would not accept the paper.

Coincidentally, a smaller study run by Swee-Lay Thein at King's College Hospital in London had measured levels of foetal haemoglobin in around 150 genetically mapped patients and had also found an association with *BCL11A*. Thein had been working for many years on the same foetal haemoglobin challenge. Her results were published in *Nature Genetics* in October 2007.[24]

'We had just got the rejection from the *New England Journal*,' recalls Sankaran. 'But [Thein] saw the same association that we were seeing.' Given my own experience of being scooped on *EGFR* (Chapter 2), I can sympathise.

But Thein's study had not included sickle cell patients, so Sankaran's work still had important academic value. It was finally published in February 2008, in the *Proceedings of the National Academy of Sciences*.[25]

Recounting the episode, Vijay Sankaran sounds sober and contemplative, but not especially upset to have been beaten to publication by another researcher. That may be because his most important contribution was yet to come.

'I realised something important that started off the real journey here,' he says. 'No one was asking, *what is BCL11A doing?* People were saying *there's this association*, full stop, that's the end of the story.'

In other words, two independent groups had found an important genetic *correlation*, but Sankaran was ready to take the next step and look for *causation*.

His goal was a simple, elegant experiment to demonstrate that *BCL11A* was indeed responsible for haemoglobin switching. He wanted to disable – or 'knock out' – the gene and see if foetal haemoglobin production resumed. Today, disabling a gene is more straightforward, but in 2008 the tools available were somewhat primitive. Sankaran used a single-stranded piece of nucleic acid with the mirror code to *BCL11A*. It would bind to any *BCL11A* RNA and either degrade it or prevent its translation.

Sankaran introduced his 'knock out' tool into some blood stem cells and waited a few days while the cells differentiated to form red cells. He then extracted the RNA from the cells and compared the quantity of adult haemoglobin RNA to foetal haemoglobin RNA.

'I got a whopping signal of foetal haemoglobin,' says Sankaran. 'It was like nothing I'd ever seen before. It was really remarkable.'

Sankaran rushed to Orkin's office. 'You've gotta see this!' he told him. Orkin grasped the significance of the result immediately. In a culture of human blood cells, disabling *BCL11A* had increased foetal haemoglobin production substantially. Sankaran had just proved the causal link between *BCL11A* and foetal haemoglobin.[26]

If the same thing could be done in a sickle cell or thalassaemia patient, it would effectively cure them. 'It was really clear that this would be a great target,' says Sankaran. 'Although *how* to target it was not clear.'

Somewhat surprisingly, though, Sankaran's part in the BCL11A story was now over. 'I wanted to go back to medical school,' he recalls, 'but it was hard to leave this.' Several mentors advised him to give up his medical aspirations and focus on research. MIT offered him the opportunity to start his own lab.

'That would have been the end of my medical career.' Ultimately, Sankaran decided to give up his cutting-edge research and become a doctor. It was, he admits, a tough decision.

Now, as a practising paediatric haematologist, Sankaran continues to research haemoglobin. He has his own lab down the hall from Stuart Orkin. He is married to another paediatric haematologist. 'I wouldn't have met her if I didn't do my medical training,' Sankaran says. He has no regrets.

Much of pharmaceutical research can be thought of as a two-step process. In the first step, we try to figure out exactly what is causing a disease by isolating one element from the immensely complex system that is the human body, and showing that it does a particular thing and that it can be manipulated with a therapy. In the second step, we go back into the body with the experimental therapeutic and try to treat the disease – safely and effectively.

For BCL11A's regulation of foetal haemoglobin switching, Vijay Sankaran had completed the first step. Now, with his departure,

Orkin needed another associate to take on the second step and show that BCL11A could be safely turned off to deliver increased foetal haemoglobin in a living creature.

Jian Xu grew up in a small village near Shanghai. He studied biochemistry at Fudan University, focusing on how genes regulate development and how gene activity can be manipulated to control cell behaviour. After gaining his PhD in molecular biology at UCLA, he moved to Boston and joined the Orkin lab in July 2008. Stuart Orkin runs his group with a light touch, allowing students and associates to choose their own projects, but his own interests permeate the lab. That's how it went with Xu. The new arrival proposed projects in the field of stem cells and epigenetics. Orkin gave his blessing, then deftly said, 'As you're interested in gene regulation, why not try a little side project on BCL11A?'

With this nudge from Orkin, Xu set out to answer two questions. Would inactivation of BCL11A deliver therapeutic benefit in a living animal? And what side effects would it cause? To answer these questions, he had to breed a lot of mice.

Xu was able to obtain a set of mice that had been bred to mimic human sickle cell disease. But mice don't have foetal haemoglobin so he had to replace their globin genes with human genes. To obtain specific characteristics in mice, researchers cross-breed varieties with different genes and then select those offspring that are closest to the result they want, then cross-breed again and again until they get to the specifications they need.

For Xu, it took eight generations to get there. Each generation was three to four months from conception to reproduction, so the whole process lasted two and a half years. The mouse colony grew so big it required hundreds of cages and took over half the room. Xu grimaces at the memory of the smell: 'When I came home, my wife would say, "Oh, you must have been working in the mouse room today."'

While he was breeding the sickle cell mice, Xu conducted safety experiments on other mice, 'knocking out' (eliminating) the *BCL11A* gene and monitoring for side effects. He knew that

BCL11A was important for brain development and the immune system, so there were serious doubts as to whether it could ever be a safe target for pharmaceutical intervention.

One way to minimise side effects was to knock out the gene only in blood cells. *BCL11A* is expressed in many cell types in our bodies, but it has different effects in different cells. Xu was able to knock out the gene in blood cells only, avoiding the brain cells and other cells where it has a vital function.

He experimented in *haematopoietic cells* – the stem cells that form all the different types of blood cells, and in *erythroid precursors* – the cells that develop into red cells.

The results were encouraging: knocking out *BCL11A* in erythroid cells increased foetal haemoglobin levels but otherwise caused very little change.[27] The gene seemed to have no – or very little – function in red cells and their precursors, other than repressing foetal haemoglobin production. 'That was great news for us,' says Xu. 'We realised this might be the target that people had been looking for over decades.'

The final proof was to come in 2010. At last, the sickle cell mice, bred over two and a half years to produce human globin, were ready. Xu knocked out the *BCL11A* gene in their erythroid cells. Then he drew blood samples.

The red cells were now a normal shape, and PCR assays showed substantially elevated levels of foetal haemoglobin. 'I could see the signal was so strong it must be real,' he remembers. 'I went straight to Stu's office – normally you make an appointment, you don't just knock on his door – and I told him, "I think this is real!"'

It was an extraordinary, historic moment. Jian Xu, with a little help and guidance from Stuart Orkin, had just cured sickle cell disease in a mouse.[28]

'It was wonderful. Fantastic,' says Orkin of Xu's achievement. 'He really solidified the work.'

Sankaran and Xu had shown – in cells and mice, respectively – that knocking out *BCL11A* led to an increase in foetal haemoglobin, which should in turn effectively cure sickle cell disease. But cutting out or disabling an entire gene in a human being, even if it were possible, would be a risky thing to do. Could we, instead, target only a small part of the gene?

Following the GWAS revelation in Sardinia, it was found that people with the SNP mutations in *BCL11A* still produced normal BCL11A protein. This implied that the mutations were not within the *exons* that are transcribed to encode the protein. Instead, they must lie in the *introns* in between. You'll remember from Chapter 1 that an intron is a region of a gene that does not encode a protein; it is the bit that is usually cut out of RNA when the exons are spliced together. Although they were once considered 'junk DNA', introns often include sequences that regulate how the genes around them are expressed.

Xu presumed, therefore, that the mutations which seemed to silence *BCL11A* in people with high foetal haemoglobin must lie in one of the gene's introns. The bit of DNA those mutations disrupted would normally act to turn on *BCL11A*. He hypothesised the existence of an *enhancer*, a piece of DNA in one of the *BCL11A* introns that would, after birth, increase the transcription of *BCL11A* and so switch off the production of foetal haemoglobin.

To find it, Xu ran an experiment comparing embryonic, foetal and adult blood cells, using a technique called epigenome mapping. He noticed that one particular intron in *BCL11A* had very strong regulatory activity *only* in adult cells. This fit with the idea of an enhancer that switched on after birth to shut down foetal haemoglobin.

Then he compared the location of this possible enhancer with the SNPs found in the Sardinia study. They matched. His biochemical analysis and the Sardinia population genetics study had pinpointed the exact same spot in one intron in *BCL11A*.

A new associate had joined the Orkin lab in 2009. Dan Bauer was another physician-scientist specialising in paediatric haematology, and he too was interested in *BCL11A*. In fact, one of his first projects in the lab involved knocking out every gene in the genome, one by one, to see if any others played a similar role to *BCL11A* in stopping foetal haemoglobin production. He had also helped Xu with the epigenome mapping study that identified the critical enhancer.

Now, as Xu turned his attention to other areas of gene regulation, Bauer took over the *BCL11A* investigation. The Sardinia group and Swee-Lay Thein had found the gene, Sankaran had confirmed it, and Xu had identified a possible enhancer within the gene. Could Bauer confirm the enhancer's role, and perhaps narrow the target down any further?

Clean-cut and youthful, with a generous, bright smile, Dan Bauer has two kids and two demanding jobs – research leader and paediatric physician – but seems calm and relaxed as he recalls his work in the Orkin Lab.

Where Sankaran had run experiments to knock out the whole *BCL11A* gene, Bauer wanted to knock out just the possible enhancer. This kind of *loss of function* study had become much easier in the years since Sankaran's work, thanks to the advent of gene editing. Now, instead of having to block the gene with a strand of nucleic acid, one could simply cut out a chunk of it.

Bauer used TALENs,* a type of programmable enzyme, to delete both copies of the possible enhancer from *BCL11A* in mouse erythroid cell lines (the cells that develop into red cells). 'The results were well beyond my expectation,' he says. 'We saw something like a ninety-seven per cent reduction of *BCL11A* activity when we deleted these enhancers.' The enhancer was clearly driving almost all *BCL11A* expression. With *BCL11A* silenced, the cells produced a lot more embryonic haemoglobin, the proxy Bauer was using in mice for foetal haemoglobin.

Working with another Orkin disciple, Elenoe 'Crew' Smith, he then did the same thing in other types of mouse cells.[29] This time, nothing happened – a wonderful result, because it meant the

* Transcription activator-like effector nucleases.

enhancer was only needed to activate *BCL11A in erythroid cells.* This meant the enhancer could be targeted in every cell in the body, and the only impact would be in the red cells where it was needed. Side effects would be minimal. This precision targeting option was an extraordinary blessing. It was becoming apparent that *BCL11A* was important to the development of blood stem cells, so targeting *all* blood cells would have caused collateral damage. Targeting an enhancer that operated only in red cells avoided that problem.

George Stamatoyannopoulos has a further role to play in this story, more than half a century after he started researching thalassaemia in rural Greece. In 2012, he attended the annual meeting of the American Society of Hematology, and there he introduced Orkin and Bauer to a molecular biologist named Fyodor Urnov.

'The only reason they even spoke with me, some random guy from a biotech company,' says Urnov drily, 'is because George introduced us appropriately.'

For a scientist, Fyodor Urnov is unusual in his tendency to draw on literary examples to make his point. His conversation is sprinkled with references to Jane Austen characters and Aldous Huxley's *Brave New World*. His parents were both professors of English Literature. 'I got beaten into submission,' he explains.

Born in the Soviet Union, Urnov was inspired to pursue a scientific career by James Watson's book, *The Double Helix*. 'DNA is the only thing worth devoting one's life to,' the young Urnov felt after reading the classic text (he has since developed a similar passion for the Beatles). 'I just can't really explain why I felt that except that this book lit that fire in me.'[30]

In 2000, Urnov joined the biotech company Sangamo, where he became one of the pioneers of gene editing. With colleagues Michael Holmes, Ed Rebar and Philip Gregory, he developed the *zinc finger* gene-editing technology and used it to great success to modify human cells in a petri dish, as well as worms, rats and agricultural plants like corn.

'We got gene editing to work,' he says with quiet pride.

Sangamo wanted to investigate how gene editing could be used in medicine, and in 2010 Urnov was asked to lead a group targeting thalassaemia and sickle cell disease. He understood the most promising solution lay in foetal haemoglobin, and for two years his group tried knocking out various genes, including *BCL11A*, to switch its production back on.

Then George Stamatoyannopoulos introduced him to Orkin and Bauer.

Urnov was being modest when he described himself as 'some random guy from a biotech company'. His role in developing zinc finger nucleases, and Sangamo's leading position in gene editing, made him just as interesting to Orkin as vice versa. At that time, Sangamo was almost the only organisation that could do gene editing efficiently and reliably.

So even though they hadn't yet published their results, Bauer and Orkin were happy to share with Urnov the precise chromosomal location of the *BCL11A* enhancer. That was all he needed.

'We bashed it with zinc fingers and we bashed it with TAL effectors,' says Urnov, referring to two early gene-editing technologies. His team, collaborating with Stamatoyannopoulos and Stam's son John, used these technologies to delete tiny fragments of DNA, one at a time, right across the *BCL11A* enhancer in blood stem cells; they then measured how much foetal haemoglobin was produced.

By October 2013, Urnov's team had found a 'soft spot' – a fragment of DNA only five base pairs long, which, when inactivated, resulted in the greatest elevation of foetal haemoglobin. It was the same result whichever technology they used. With two independent approaches yielding one identical answer, they knew they had the key to the enhancer – a kill switch that would inactivate it.[31] They designated those five DNA base pairs the gene's 'Achilles heel'.

'I can comfortably count the number of times Mother Nature

has revealed her secrets to me on the fingers of one hand,' says Urnov. 'I don't think I will ever forget that moment.'

In parallel, in the Orkin lab, Dan Bauer was zeroing in on the same scrap of DNA.

The previous year, gene editing had reached a whole new level of sophistication with the discovery of *CRISPR/Cas9* by Emmanuelle Charpentier and Jennifer Doudna. They and others had turned a natural bacterial defence against viruses into a remarkable tool for making precise cuts at specific points in our DNA, enabling researchers to insert or delete sequences in human genomes with ease. We'll come back to this shortly. Dan Bauer instantly saw the potential value of the technology in his work on the *BCL11A* enhancer, and he reached out to Feng Zhang, a pioneer in the development of CRISPR/Cas9 as a genome-editing tool.

With Zhang's help, Bauer and MD PhD student Matt Canver used CRISPR/Cas9 to delete short sequences right along the length of the enhancer DNA, just as Urnov was doing with zinc fingers and TAL effectors. By methodically working their way along the enhancer with a CRISPR scalpel, Bauer and Canver were able to identify the same five base pairs as the lynchpin to foetal haemoglobin production.* [32]

This tiny fragment of DNA, we now know, is the location on the *BCL11A* gene that binds the GATA-1 transcription factor that Stuart Orkin discovered in 1989. GATA-1, remember, is the protein that regulates much of what happens in red cells. And so, from the work of Orkin, Sankaran, Xu, Bauer, Urnov, Stamatoyannopoulos and their colleagues, we now have the whole picture: in erythroid cells, GATA-1 binds to these five base pairs, causing the enhancer to activate *BCL11A*, which switches off the production of foetal haemoglobin. If we can knock out this Achilles heel, there is nothing for GATA-1 to bind to and *BCL11A* remains inactive, allowing foetal haemoglobin production to continue.

* In fact, Bauer and Canver found three vulnerable sites within the enhancer, of which the most sensitive was the same Achilles heel identified by Sangamo.

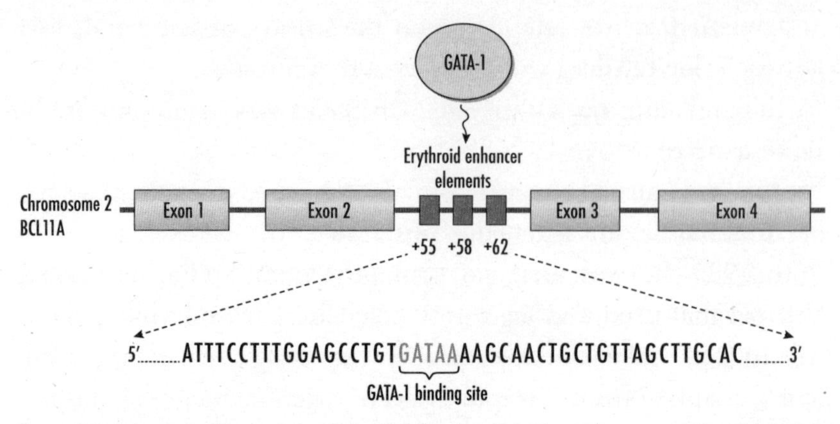

Figure 20. Schematic illustration of the 'Achilles heel' in the enhancer region of BCL11A.

Where Sankaran had resurrected foetal haemoglobin by knocking out expression of an entire gene, and Xu had discovered an erythroid-specific enhancer within one intron in that gene, Urnov and Bauer had separately found just five DNA base pairs within that enhancer that acted like a kill switch. They had narrowed down the therapeutic target to something unimaginably small and precise.

And with CRISPR/Cas9, the world now had the perfect technology to target that Achilles heel in sickle cell and thalassaemia patients.

Step out of Boston Children's Hospital, where Orkin, Sankaran, Xu and Bauer were cracking the secrets of BCL11A, and it's an easy walk to a number of other famous institutions. Wander up Brookline Avenue and you quickly reach Fenway Park, home of the Red Sox. Stroll onward to the Charles River, cross Harvard Bridge, and you come to the Massachusetts Institute of Technology. MIT is the home of the Broad Institute, and it was here that geneticist and physician David Altshuler first heard about the work being done in the Orkin lab.

'Stu and Vijay and Dan really just did this extraordinary job of figuring out the biology of BCL11A,' he says. 'In an absolutely critical set of studies, they figured out over a period of years exactly how this worked.'

It wasn't the first time Altshuler had contemplated a potential cure for sickle cell disease. As an undergraduate interested in molecular biology and medicine, he crossed paths with Vernon Ingram, the researcher who in 1956 had discovered the genetic cause of the disease. 'You learned about sickle cell if you went to MIT, because it was, like, *the* canonical genetic disease,' says Altshuler.

In 1985, he was assigned to a project at MIT's Whitehead Institute that used a virus to insert genetic code into blood stem cells to treat sickle cell disease. It was a very early attempt to use gene therapy to combat the disease.

'This was not a new idea,' insists Altshuler. 'This, I had been thinking about for thirty years.'

As an intern at Massachusetts General Hospital in 1995, Altshuler was working in the intensive care unit when a twenty-year-old African student with sickle cell disease was admitted with terrible bone pain. 'It was a sunny day. He seemed to be doing well. He was in pain, but he seemed to be doing well. We were then called to his bedside because he'd coded [suffered a cardiopulmonary arrest]. His heart had stopped. He died. We couldn't understand what happened. We couldn't get him back.' An autopsy later revealed the patient had had a fat embolism: the sickle cell disease had caused one of his bones to die, and the fatty marrow from it had got into the bloodstream and blocked his lungs.

Altshuler leans back, his arms tightly crossed and face drawn, as he recalls the tragic loss. 'He died . . . and I was so upset. He was an MIT student like me, a lovely guy. He died and he had a fat embolism and I remember being deeply upset. He came in 9 a.m., a healthy-looking person, and died two hours later.'

David Altshuler is a poised, assured man, his confidence built on decades of success in academia and then pharma. Behind him in his office is a Swiss exercise ball, on which is balanced what appears to be a doll or puppet modelled on himself – same goatee beard and calm gaze, finished off with a white lab coat. On the floor beside it are two bottles of champagne. Not the average Zoom background.

He is full of praise for the work done by the Orkin lab, but adamant that their achievements did not mean the battle against sickle cell disease had been won: 'It's really easy to treat cells in a dish; it's easy to cure mice. It's been done thousands of times.' Pulling off the same trick in humans, he insists, is much more difficult. 'If you talk to people who really excelled in academia and then go to try and make medicines for people, it's, like, a *very hard thing*.'

Anyone who supposed the job was done when the fundamental science was described would be swiftly put right by Altshuler. 'It would be like saying you've built a bridge of Lego, and now you think you know how to build a bridge across the Hudson.'

He has a point, and it echoes a broader lesson for innovators. Developing a functioning prototype in a lab is great, but designing and delivering an innovation that works at scale across the globe often requires another level of ingenuity and resources.

While working at the Broad Institute, Altshuler served on the board of a young pharmaceutical company called Vertex. In 2015, he was invited to become its Chief Scientific Officer and develop a new R&D strategy for the company. At the time, Vertex had no interest in sickle cell disease. The company had made a name for itself creating the first medicine that treated the underlying cause of another rare genetic disease, cystic fibrosis. They were also working on cancer, rheumatoid arthritis and flu.

With the agreement of the new CEO, Altshuler terminated or sold almost all Vertex's existing research programmes, a remarkably bold move, and initiated a new sickle cell disease and thalassaemia programme, drawing on the Orkin lab findings and making use of the new CRISPR/Cas9 gene editing technology.

'What was clear to me was *this* was the project,' says Altshuler.

It would be easier than some gene-editing projects because blood cells could be removed from the body, edited *ex vivo*, and then reinserted into the patient. Some diseases would require gene editing *in vivo* – inside the body – a much harder and riskier prospect. But Vertex would not have the field to themselves: Sangamo, the California biotech that had identified the Achilles heel, was already working on their own *BCL11A* enhancer gene-editing programme using zinc fingers to target the same two diseases.

Lacking in-house CRISPR/Cas9 expertise, Vertex partnered with CRISPR Therapeutics, the Swiss-based company set up in association with Emmanuelle Charpentier. Since its formation in 2013, CRISPR Therapeutics had prioritised sickle cell disease and thalassaemia as therapeutic targets for the new technology. One of its scientific founders, Matthew Porteus, had previously worked on elevating foetal haemoglobin using different gene editing tools. Indeed, CRISPR Therapeutics had already identified five strategies, editing different chromosomes, that could potentially lead to elevated foetal haemoglobin levels.

By the time Vertex got in touch in 2015, CRISPR Therapeutics had achieved results in cell lines and mouse models that demonstrated the elevation of foetal haemoglobin in haematopoietic stem cells. To take the next step towards a viable therapy, they welcomed the offer of a collaboration. But rather than a traditional partnership, with a large pharma company licensing the technology and developing the product, CRISPR Therapeutics was much more interested in a co-development partnership model that would combine the unique strengths of the two companies.

That suited David Altshuler just fine. 'I'd come from the Broad Institute where everything was collaborative,' he explains. 'I didn't know how things worked in companies – you know, give people money and wait and see what happens – so we did in fact create a joint project team.'

'When we were striking the partnership with Vertex, it was very important to me that it was an equal partnership,' says Samarth Kulkarni, now the CEO of CRISPR Therapeutics. 'Because it would bring the best of both worlds – the nimbleness and specific expertise of a smaller company and the established and experienced drug development team of a larger company. There was a wonderful spirit of collaboration.'

The joint project team had first tried directly editing the single-letter mutation in the *HBB* gene that causes sickle cell disease. However, correcting a single-letter mutation is still very difficult to do safely, even with CRISPR/Cas9. There is a risk of deleting an adjacent bit of code by mistake, disrupting this important globin gene and so actually *causing* thalassaemia. Letter-by-letter gene editing is a work-in-progress.

To correct the fault in the *HBB* gene would be like inserting a tiny missing spring into a highly complex Swiss watch, without disrupting any of the other mechanisms. By contrast, to disable the *BCL11A* enhancer, all they needed to do was cut out the Achilles heel. In life, it's generally easier to break things than fix things; the magic of the BCL11A enhancer is that it allows us to fix the disease by breaking something. And because the enhancer is erythroid-specific, breaking the Achilles heel doesn't affect anything other than red cells.

The two companies divided up the work according to their competences, with CRISPR Therapeutics doing most of the discovery work to create the CRISPR/Cas9 editing tools and figure out the manufacturing processes, while Vertex led on the later process refinement and clinical trials.

Chad Cowan and Rodger Novak assembled a CRISPR project team of around twenty people. In Pharma, that's not a big number, but for CRISPR Therapeutics that was most of the workforce, plus a few consultants. This one project constituted about ninety per cent of the young company's entire investment. They were effectively betting the house on *BCL11A*.

This is complex medicine, so let's start simply. The thing Vertex and CRISPR Therapeutics have made is not a pill or a capsule. 'For all practical purposes, you should think of our product as a bone marrow transplant,' says Bastiano Sanna, now head of cell and gene therapy at Vertex.

The bone marrow holds the blood stem cells that develop into red cells. Change the genes in those stem cells, and you may effectively cure thalassaemia or sickle cell disease.

Thalassaemia can, in rare cases, be treated by a bone marrow transplant from a healthy person. This procedure is very difficult and risky. One complication that can arise is rejection of the donor's cells, so the patient's immune system has to be suppressed.

With the gene editing approach, the transplant uses the patient's *own* cells, so there is minimal risk of the immune system attacking them. Blood stem cells are extracted from the bone marrow, the genes in them are edited, and then they are reinserted into the patient.

The first step in this process, the extraction, is now straightforward, thanks to the development of mobilisation agents – drugs that cause blood stem cells to come loose from the bone marrow and circulate in the bloodstream, where they can be harvested from a vein. This is a very welcome innovation. Previously, says sickle cell physician Bill Hobbs, 'we would stick a big, giant needle into the hip bones and suck out enormous amounts of bone marrow. You'd be doing 15–20 needle punctures on each side of the pelvic bone to get sufficient bone marrow stem cells.'

The last step is also fairly straightforward. Edited blood stem cells injected into the bloodstream will find their own way back to the bone marrow. The complication here is that to make room for the new, edited cells, we have to eliminate the old bone marrow cells with chemotherapy. This is a risky step that must be carried out in a specialist transplant unit. While the stem cells are self-replicating and reconstituting the bone marrow, the patient's immune system is largely out of action, and so the patient has to remain isolated and under close supervision.

That leaves the middle step – the gene editing itself.

The development of CRISPR/Cas9 gene editing is a wonderful story that is worth reading in full.[33] To summarise briefly, certain bacteria have evolved a system to fight viral infection. A strand of the bacteria's RNA – called *guide RNA* – carries a genetic sequence that matches part of the virus's genetic code.* You can think of this as the photo on a police *Wanted* poster: it identifies the villain. The guide RNA binds to the viral DNA, guiding an enzyme called *Cas9* to cut the viral genome, thus disabling the virus. Cas9 is the SWAT team, in this analogy, called in to eliminate the villain. Charpentier, Doudna and others realised that this precision DNA cutting tool could be used to make cuts in *any specific* DNA sequence (including human DNA) if the guide RNA could be modified to carry the matching sequence. So CRISPR/Cas9 gene editing was born.

At a conceptual level, the process sounds relatively simple: all that is needed is a Cas9 enzyme and a strand of guide RNA containing a fragment of code matching the target DNA. In this case, of course, the target is the Achilles heel in the BCL11A enhancer. To disable *BCL11A*, we need a strand of guide RNA that recognises those five base pairs in the Achilles heel and guides Cas9 to cut them out. The cell nucleus will then repair the DNA without the Achilles heel, leaving an inactive *BCL11A* gene. All future blood stem cells and red cells grown from those cells will carry the same inactive gene.

The CRISPR Therapeutics team tested thousands of possible guide RNAs and Cas9 enzymes in the laboratory to find the safest and most drug-like candidates. They were able to draw on work by academic groups, including the Orkin lab, which had identified and published several guide RNAs, but these had to be painstakingly assessed and characterised in multiple cell types and animal models before they could be considered for a human trial. But that was only the start of the challenge.

As Altshuler makes clear, the process was very far from simple when it came to treating human beings. 'The process *is* the product,' he explains.

* Parts of the guide RNA sequence are repeated over and over again in the bacterial genome, giving rise to the slightly forced CRISPR acronym: Clustered Regularly Interspaced Short Palindromic Repeats.

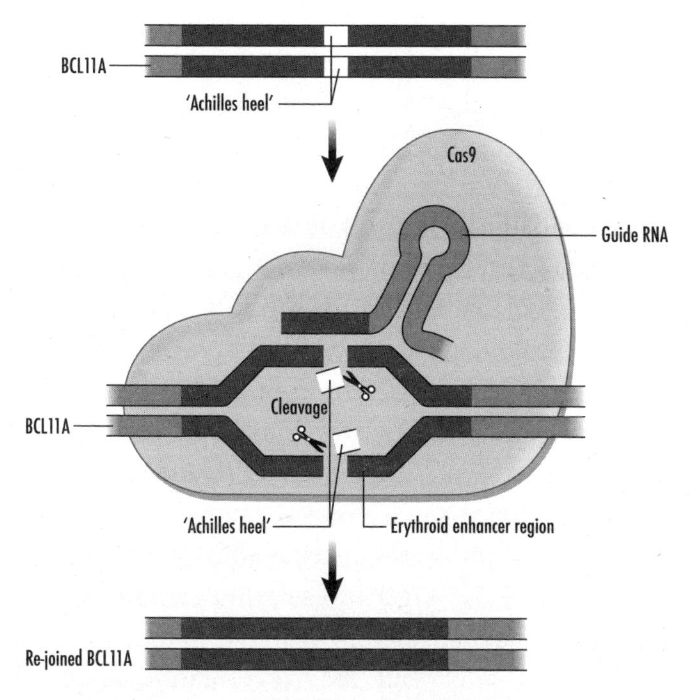

Figure 21. The 'Achilles' is cut out of BCL11A by CRISPR/Cas9.

How, for example, should we go about getting the guide RNA and Cas9 enzyme into the patient's cells? One method is electroporation: we apply an electric shock to extracted blood cells, to briefly open their pores and allow entry of new genetic material. But this is very hard to do at scale. The shock can easily kill the cells. It took the joint team many months to get the process right.

'It's a large engineering problem, with expertise needed across cell biology, bioinformatics, molecular biology and biophysics, to make one of the most sophisticated medicines ever developed,' says Kulkarni.

Then they had to work out a robust process to grow billions of blood stem cells from the edited cells without compromising their ability to self-reproduce. Only a tiny proportion of blood stem cells are capable of reproducing themselves, and this vital *long-term* lineage is easily lost in the cell culturing process.

The Vertex-CRISPR team experimented with multiple approaches to ensure long-term cells survived the process. They conducted sixteen-week animal transplants of the edited cells for each candidate treatment approach, to check that long-term cells had been successfully edited and transplanted. Through trial and error, they learned how best to culture the cells to ensure long-term reconstitution.

A key choice early on was to develop robust, scalable processes from the start, rather than rushing to prove the therapy worked in a couple of patients and then backward engineering more sustainable processes. This strategy was viable, because Vertex and CRISPR Therapeutics were secure in their financing and confident of their ultimate success; other innovators might have preferred to cobble together a proof-of-concept with simpler tools and processes, then make the larger investment in long-term processes once they had shown their therapy worked.

When asked to single out the greatest innovative achievement of the joint project team, Bastiano Sanna gives a surprising answer. 'The biggest challenge for this kind of therapy is actually a boring one: consistency.' Every patient is different, and the raw materials needed for the therapy – their cells – can behave unpredictably. So the team had to create a process that was robust and flexible enough to accommodate the variability in cell types of the many different patients they hope to treat. The result was a very lengthy process with many steps, each of which had to be optimised. 'It's multiple layers of incremental innovation.'

Fyodor Urnov agrees. 'The consistency of making the cells is astonishing and impressive. This is a huge, huge win for the CMC [Chemistry, Manufacturing and Controls] team at CRISPR Therapeutics. I really salute them for it. It's very tedious work, and hard work, and requires the right mindset, and the right hands. It's artisanal. These people are incredibly skilled in a very specialised way and I admire them tremendously for it.'

After a long period of testing in animal models, the joint team finally had a Cas9 enzyme, an RNA guide, and a treatment process – a therapy package collectively termed CTX001.*

By late 2018, the Vertex-CRISPR team were ready for clinical trials. Sangamo had beaten them to this critical first-in-human milestone with their rival zinc-finger-based BCL11A enhancer gene-editing process, but Vertex-CRISPR were the first team in history to open an Investigational New Drug application with the FDA to treat human beings using CRISPR/Cas9. Unsurprisingly, the FDA had a lot of questions, and they invited the two companies to Washington DC to go over their trial proposal.

For three days, the joint team holed up in a DC hotel, practising their presentation and figuring out answers to any possible question the FDA might ask. Their extensive preparation was time well spent. When the day of the meeting arrived, numerous officials piled into the conference room, fascinated by the new technology. They interrogated every aspect of the process Vertex and CRISPR Therapeutics had developed, probing for any possible risk to patients. By the end of the day, everyone was exhausted.

But the FDA was satisfied, granting permission for the first ever human trials of CRISPR/Cas9 gene editing.

The first patient was a young woman in Germany. Approximately one billion of her blood stem cells were harvested, refrigerated and flown to the Roslin Institute in Scotland, the contract manufacturer that performed the gene editing. There, they were incubated in culture media and treated with CRISPR-Cas9 gene-editing reagents. Following a recovery period in culture media, the cells underwent rigorous quality testing and were then flown back to Germany for re-infusion into the patient. Subsequent patients' cells had to make similarly complex international journeys.

The results, in the handful of patients so far treated, have been impressive.

* Now named exagamglogene autotemcel (exa-cel).

'I was shocked when I saw the clinical data,' says Bill Hobbs, who joined Vertex in 2020 to lead the clinical development of CTX001. 'It exceeded my expectations of the amount of foetal haemoglobin that would be produced by targeting BCL11A in people.' The effect is, he says, 'amazingly reproducible', which for scientists is the gold standard evidence that something is true or that a medical process works. 'All of the [sickle cell disease] patients have a very similar trajectory of increasing foetal haemoglobin. By about three to four months, they're at about thirty per cent; after that they're about forty per cent. The variability in foetal haemoglobin is incredibly tight.'

Victoria Gray was the first sickle cell patient treated in the United States. For her, CTX001 has been life-changing. 'It's amazing,' she told NPR. The extreme pain she had suffered for much of her life is gone. She has dispensed with the powerful pain medications she depended on for decades. 'It's better than I could have imagined. I feel like I can do what I want now.'[34]

Foetal haemoglobin levels in her blood remained high, and bone marrow samples taken a year after treatment showed the gene-edited cells still present: her DNA seemed to have been permanently altered.

'This gives us great confidence that this can be a one-time therapy that can potentially be a cure for life,' says Kulkarni.

At the time of writing, CTX001 is still in clinical trials, but the results are encouraging.[35] Eighty-three patients – forty-eight with beta thalassaemia and thirty-five with sickle cell disease – have been treated. The vast majority have responded well, with the main significant side effects resulting from the chemotherapy needed to clear their bone marrow. Responding patients are synthesising much higher levels of foetal haemoglobin and their symptoms have receded. All but two of the thalassaemia patients no longer need blood transfusions.[36]

Such success is still unusual in the nascent field of cell and gene therapy. 'Nothing ever works and everything fails all the time,' says Fyodor Urnov. 'So in those rare cases where something works, people should just all together jump into a swimming pool of champagne.'*

Gene therapy is truly amazing. The idea that Vertex and CRISPR Therapeutics can edit someone's blood stem cell genes and potentially cure an inherited disease seems like fantasy. Theirs is the first ever practical application of CRISPR/Cas9 gene editing in medicine. It's a magnificent achievement. But there is one big problem. As you might have guessed while reading about the complicated, highly skilled process and the weeks patients spend in hospital, it is phenomenally expensive.

To treat a single patient could cost in excess of two million dollars.

Even if the process becomes substantially cheaper, we can be certain this is not a solution that will be available to most of the millions of sickle cell patients in Africa or thalassaemia patients in Asia.† CTX001 is a treatment that only a small proportion of patients will be lucky enough to afford.

But this isn't the end of the road. The gene-editing technology, says David Altshuler, is not the point:

'One way of thinking about what we've done with CTX001 is that we turned on foetal haemoglobin in erythroid cells and showed it could treat the disease. The question is, now, what's the simplest, safest and most scalable way to do that?'

To Altshuler, then, CTX001 is merely the proof of concept for George Stamatoyannopoulos's hypothesis: if we can reverse the haemoglobin switch and bring back foetal haemoglobin, we can treat sickle cell disease and thalassaemia. Now we have to find the best way to do it.

* Urnov's former employer, Sangamo, also had some success with gene editing *BCL11A* in clinical trials, using zinc fingers rather than CRISPR/Cas9, but they ultimately terminated the project.

† As well as the cost barrier, there is a facilities barrier: the procedure needs to take place in a specialist transplant unit, and they are few and far between in Africa and large parts of Asia.

Bill Hobbs agrees. 'The first iteration of a new technology is never where you end up. CTX001 is great, and has the potential to help a lot of people, but our work is not done yet.'

One possibility is to find a way to initiate gene editing in the body. Instead of having to take bone marrow cells out of the body and destroy those left behind, perhaps we could inject a gene-editing agent into the bone marrow, and let it slowly convert all the cells *in situ*. 'That's science fiction today,' acknowledges Altshuler, 'but the whole thing was science fiction ten years ago, so . . .' CRISPR Therapeutics has in fact already embarked on an *in vivo* gene editing project.

An even better answer would be a small molecule – a drug cheap enough and simple enough to offer to every sickle cell disease and beta thalassaemia patient in the world. But it won't be easy to discover. If we target the same switching mechanism, a small molecule would have to block activation of *BCL11A* in red cells without impacting the gene in other parts of the body where it's needed. And to date, no small molecules have been found that block a transcription factor like GATA-1.

Vertex has had forty people working for five years on a project to develop a small molecule that will one day render their gene-editing triumph redundant. 'One of the things that keeps me coming into work every day is the dream that someday we will have a pill that can raise foetal haemoglobin,' says Altshuler.

Stuart Orkin has the same ambition. 'We've been working on it for a long time,' he laughs. 'It's an exceedingly difficult project.' He doesn't have the same resources as Vertex, but he's proved his dedication to the foetal haemoglobin quest over decades, as well as his ability to hire the right people to pursue it. So, will he manage to invent a pill for sickle cell disease and beta thalassaemia before he permanently goes to Hawaii? It's a long shot, perhaps, but his priorities are clear: 'If you ask me what I get up in the morning to do, this is it. This has potential impact that is much greater than anything else that I could possibly do in the laboratory.'

CHAPTER 8

Viral Geometry

Condition: HIV/AIDS
Target: HIV capsid
Innovation: Lenacapavir [len-ah-CAP-ah-veer]
Company: Gilead Sciences

A NEW VIRUS SPREADS WITH TERRIFYING speed all over the world, sickening and killing people everywhere . . . We all know this story only too well. We hope and pray it will not happen again in our lifetimes. But those of us over a certain age have witnessed this tragedy before. That pandemic was far more deadly, claiming an estimated forty million lives and counting. Back then, it took us much longer to figure out what it was and how to combat it with drugs. We've made a lot of progress, but the hunt for treatments still continues more than four decades later.

When five otherwise healthy young men in Los Angeles were diagnosed with pneumocystis pneumonia in June 1981, their physicians were perplexed.[1] This rare fungal infection of the lungs is usually harmless in humans, unless the immune system is substantially compromised. The patients had no history of immune deficiency, so why were they so sick? Had they somehow *acquired* an immune deficiency? Doctors had never seen such a thing before.

Further cases soon emerged in San Francisco and New York.

Another unusual symptom was showing up: Kaposi's sarcoma, a rare skin cancer that manifests as striking purple patches all over the body. 'At that point,' recalled Anthony Fauci, the former director of the National Institute of Allergy and Infectious Diseases, 'it was like, "Oh my God, we're dealing with something new."'[2]

Many of the patients were men who had sex with men (MSM).* They developed enlarged lymph nodes all over their bodies.† Their conditions rapidly worsened, leading to extreme weight loss, multiple organ failure and death.

At first, it was rumoured that the strange new illness was caused by recreational drugs, such as cocaine or amyl nitrate. But soon it was found to be sexually transmitted. By the following year, intravenous drug users and people with haemophilia were also developing the same fatal symptoms, a strong indication that they were caused by some kind of blood-borne pathogen. Cases also started to appear in heterosexual people, and in infants, suggesting that mother-to-child transmission was possible. The mysterious condition was named *Acquired Immunodeficiency Syndrome* (AIDS) in July 1982.

The coming scale of the epidemic was increasingly apparent. Cases were multiplying at an exponential rate. Afraid and largely ignorant about the disease, the rest of society subjected infected people and their partners to appalling prejudice. The discrimination extended to people with haemophilia: in a few awful cases, children with haemophilia were ostracised by their friends and excluded from school.

Activists took the lead on infection prevention by promoting safer sex and spreading awareness about how the disease was transmitted. But it was clear a pharmaceutical response was also needed.

That became slightly more feasible in 1983, when Françoise

* We use this term in epidemiology, rather than 'gay men', because not all gay men are sexually active and not all MSM identify as gay.

† Lymph nodes are a key part of the immune system.

Barré-Sinoussi and Luc Montagnier first sighted the retrovirus now known to be the causative agent of AIDS.[3] They cultured lymph node cells from a man with swollen lymph nodes and detected *reverse transcriptase* activity in the growth medium. Reverse transcriptase is an enzyme characteristic of retroviruses – the class of viruses that can integrate some of their genetic code into their host cell's DNA. On closer examination with an electron microscope, they found retroviral particles – new viral molecules – forming on an infected cell. When isolated, Barré-Sinoussi and Montagnier showed that these previously unknown viral particles could infect and kill healthy lymphocytes, which are central to our immune system.

Soon thereafter, an international committee named the named the new pathogen *Human Immunodeficiency Virus* (HIV).

Identifying the virus made a diagnostic test possible. Testing revealed thousands of seemingly healthy people around the world were already infected. Indeed, it showed that HIV had been spreading through the human race for decades. HIV type 1 (HIV-1), group M, the most common type among AIDS patients, is now understood to have originated from a chimpanzee virus.* The earliest specimen of HIV-1 that we have identified among historic blood samples was collected in 1959, in what is now the Democratic Republic of the Congo.[4]

The first breakthrough against HIV was an old, failed cancer drug called azidothymidine (AZT). It was discovered in the 1960s, tested in mice and found not to work against tumour cells. Shelved for two decades, it was resurrected as one of many candidates for a mass screening of potential antivirals by British pharma company Burroughs Wellcome.† [5]

No silver bullet, AZT did nonetheless help to slow the progression of AIDS in many patients. But it wasn't long before the remarkable ability of HIV to mutate and evolve undermined its

* This chapter is concerned only with HIV-1.

† This was made possible by HIV pioneer Robert Gallo, who developed a system for HIV growth in cell culture that enabled all the subsequent drug screening.

therapeutic benefit. HIV can change its genome so rapidly to evade medication that patients on AZT got just a few months of respite before virus levels began to rise again.

New therapies, designed to target HIV's particular characteristics, had to be developed.

Wes Sundquist was at college in Minnesota, his home state, when the AIDS epidemic erupted. His subject was chemistry, which he pursued with a PhD at MIT. By the late 1980s, he had developed an interest in structural biology, and he took a postdoc position at the Medical Research Council Laboratory of Molecular Biology in Cambridge, UK.

Like many people at the time, Sundquist held the basketball star Magic Johnson in high regard. 'He was an icon,' he says. When Johnson announced in 1991 that he was HIV-positive, the news made a huge impression on Sundquist. He'd been as concerned as everyone else about the global epidemic, but that announcement somehow threw the magnitude of the problem into sharper relief. When one of his colleagues started working on the virus, Sundquist became curious about HIV as a research subject. He had been offered a position at the University of Utah, and the chance to set up his own research lab, but had not yet found a topic that engaged him. Could this be it?

By then, lots of labs and pharma companies were piling into the HIV field. It was one of the biggest health emergencies on the planet. Most of their research was focused on the obvious drug targets: the critical enzymes that enabled viral replication, such as the protease (see Chapter 6) and reverse transcriptase. Other researchers were looking at the spike proteins that allowed the virus to enter cells. With so many people already working on those targets, Sundquist decided to find a different aspect of HIV to study.

Each HIV particle, or *virion*, consists of a *viral envelope*, which encloses a protein shell called a *capsid*; the capsid holds two strands of RNA and the enzymes needed for replication. Picture a peach or a cherry: the viral envelope is the soft skin of the fruit; the capsid is the hard shell of the stone/pit; the RNA and enzymes make up the seed inside.

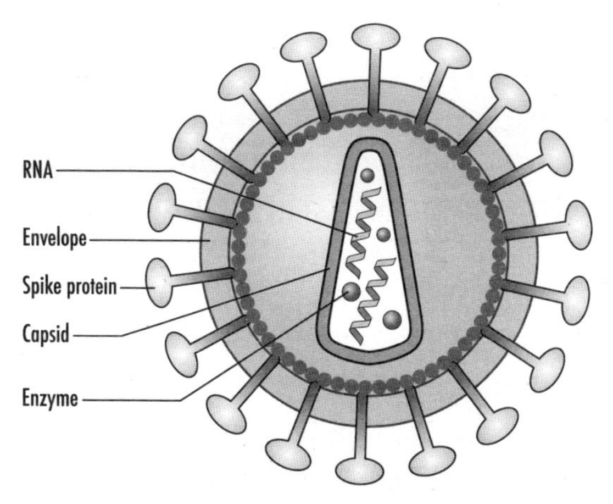

Figure 22. Structure of an HIV particle.

Drug hunters had largely ignored the capsid. Its hard lattice of interlocking structural proteins looked impenetrable. Drugs usually work by attaching to some existing binding site, which is why enzymes make such good targets – the whole point of an enzyme is to bind to other molecules and drive some kind of change. The function of the capsid, on the other hand, seemed all about containment and protection. Why would a protective shell have binding sites?

But Sundquist wasn't a drug hunter. He was not really thinking at that point about HIV treatments. What grabbed his attention was the geometry of the capsid.

It was a cone.

'There was no known cone in biology at that time,' he explains.* 'Honestly, the thing that I was most interested in was the fact that it was a cone.'

The early electron microscope images of HIV particles didn't show much detail, but that unusual cone was a striking feature. No

* Pine cones don't have a continuous conical surface.

one knew much about it, other than what it was made of: a few thousand strands of p24 **ca**psid protein, or *CA protein*. Sundquist initially studied the structure of CA protein, which he managed to describe in collaboration with two other researchers, Mike Summers and Chris Hill.[6]

Buoyed by that early success, Sundquist and his colleagues set out to establish the basic structure of the capsid and how it is constructed. They found evidence that simple strands of CA protein, *monomers*, could come together to form *hexamers*, protein structures made up of six monomers.[7] The hexamers took the form of hexagons (six-sided rings). This seemed to be the basic unit of construction, the bricks that made up the capsid wall. But Sundquist had studied the theorems of Leonhard Euler, the eighteenth-century Swiss mathematician, and he knew that an enclosed space (like a sphere or a cone) could be formed out of hexagons only if the structure also included at least twelve pentagons (five-sided rings). You can see this on a soccer ball, which is made up of twenty hexagons and twelve pentagons.

However many hexamers a capsid might contain, Sundquist reasoned, it must also have at least twelve *pentamers*.

From elemental carbon science, he borrowed another theory: that we can make a cone out of hexamers and pentamers if we arrange seven pentamers at the wide end and five at the narrow end.

With these two ideas underpinning his hypothetical model of its structure, Sundquist was able eventually to show that the HIV capsid is a cone made up of approximately 250 hexamers and exactly twelve pentamers.

Next, Sundquist and his team wanted to understand what the capsid did. You might think that was obvious: it looked like a robust shell enclosing the virus's genetic material, so its function was clearly containment and protection. Yes, but . . .

Sundquist wanted to know if the capsid played other roles. To assess its function, his team made a number of small mutations in the capsid's genes. This was only possible because a new researcher had joined his lab: Uta von Schwedler had extensive experience

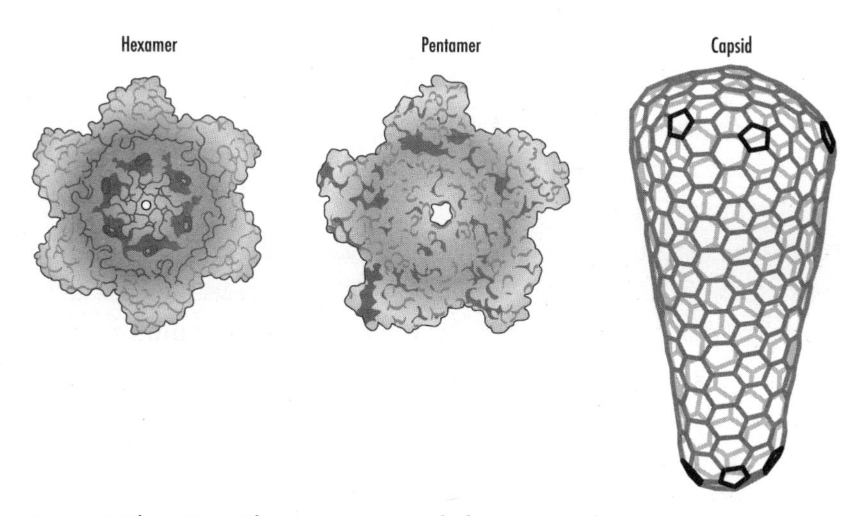

Figure 23. The HIV capsid is a cone comprised of approximately 250 hexamers and exactly 12 pentamers.

growing HIV in a safe and secure laboratory. Previously, Sundquist had worked only with CA protein; he had never handled the live virus. With von Schwedler's expertise, the team could now safely engineer new strains of HIV and test how well they replicated and infected new host cells.

The mutations were designed not to disrupt the structure of the capsid, and indeed many of the resulting genetically modified strains of HIV had seemingly intact capsids. But almost all the mutations rendered the virus much less infectious. In fact, of the forty-eight point mutations they engineered, twenty-seven caused a twenty-fold decrease in infectivity: the capsid often appeared intact, and yet the virus was really struggling to replicate.

These mutations proved that the capsid played a really important part in the HIV life cycle, beyond mere structural containment and protection.[8]

By the mid-1990s, the AIDS epidemic had become a pandemic, with over three million new HIV infections and more than a million AIDS-related deaths each year.[9] A million children, most of them in sub-Saharan Africa, had been orphaned by AIDS.[10] The numbers just kept going up and up.

In December 1995, the first protease inhibitor was approved by the FDA. Saquinavir was developed by Roche and proved to be a game-changer in the fight against HIV. A cocktail of one protease inhibitor and two nucleoside reverse transcriptase inhibitors (NRTIs) was shown in trials to reduce viral loads to very low levels for up to a year. When these findings were announced at a conference in Vancouver, attendees gave a standing ovation, hugged each other and cried. This so-called highly active antiretroviral therapy, or HAART, would be a lifesaver for thousands.[11]

But those early regimes were by no means ideal. People with HIV had to take up to thirty tablets a day, at specific times. Missing scheduled doses could mean drug resistance and therapy failure. The debilitating side effects included round-the-clock nausea and diarrhoea. Often the medication would significantly change the shape of a person's face and upper body, due to the loss or redistribution of fat under the skin, telegraphing their disease to the world. Many found these outcomes so intolerable they would give up their lifesaving medicine: nearly a quarter of patients discontinued HAART within the first eight months.[12]

For those who persisted with treatment, drug resistance remained a looming threat that might render their therapy ineffective at any moment.* Then, there was the cost of the drugs, making them inaccessible to millions of people in less wealthy nations. By the start of the new millennium, as Wes Sundquist was preparing to publish his findings on the capsid, there was still huge unmet need in the treatment of HIV and AIDS.

Tomas Cihlar got interested in viruses at an early age. 'I was fascinated with the biology world, and viruses in general,' says Cihlar.

* The development of other antiretrovirals, including the non-nucleoside reverse transcriptase inhibitors (NNRTIs), helped to slow the onset of resistance.

'They're perfect small parasites, hijacking perfectly healthy cells and turning them upside down to work for the virus.'

His biochemistry PhD brought him into the orbit of an antiviral legend, Antonín Holý, at the Institute of Organic Chemistry and Biochemistry (IOCB) in Prague. Holý was working on molecules to counter the herpes virus, and he invited Cihlar to join his lab.

Holý was a key player in a highly productive partnership between the IOCB, the Rega Institute in Belgium, and a small Californian biotech start-up called Gilead Sciences. Founded just a few years earlier, Gilead did not yet have any commercial products, but they had investors and they had vision. Through Holý's friendship with Gilead's head of R&D, John Martin, the young Cihlar found himself offered a postdoctoral fellowship in California. 'I kind of sneaked through the back door,' he says, smiling.

It was an invigorating research environment. Martin was determined to build a pharmaceutical powerhouse to take on viral diseases, including the great threat of the age, HIV. Cihlar initially worked on cidofovir, a herpes drug discovered by Holý that became Gilead's first approved commercial product in 1996. But it wasn't long before he was drawn to the company's HIV programme.

'Tomas is a spectacular person,' says John Link, a medicinal chemist. 'He's an incredibly unusual combination of a drug discoverer and a very academic thinker. They're usually two separate universes.' Although Cihlar can come across as stoic and businesslike, Link knows him better: 'He's got a great sense of humour: very wry.'

The big breakthrough, for both Gilead and Cihlar, came in 2001, when tenofovir disoproxil fumarate (TDF) was approved for use against HIV. At a time when resistance was mounting against all existing HIV drugs, TDF became a lifeline for many patients. 'That was the seed of the whole Gilead HIV portfolio,' says Cihlar.

Gilead combined TDF with another antiviral, emtricitabine, giving us one of the most important HIV treatments so far created. The combination is still prescribed to millions of people. Its approval in 2004 set Gilead on course to become the leading player in HIV therapeutics. And that's when Cihlar, by now Gilead's leading virologist, started wondering about new potential targets in the virus.

The drugs available to people with HIV had turned the disease from a death sentence into a chronic condition, but many of them still had potential side effects, and the threat of drug resistance hovered over all of them. There were the protease inhibitors, the nucleoside reverse transcriptase inhibitors and the non-nucleoside reverse transcriptase inhibitors. All of them targeted HIV enzymes.

Was there a completely different target they could try?

In 2005, Cihlar came across one of Wes Sundquist's papers. Written four years earlier, it described the mutation studies that showed how small alterations in the HIV capsid massively diminished the infectivity of the virus. The mutants created by Sundquist and von Schwedler had been further investigated by another capsid expert, Christopher Aiken of Vanderbilt University, who developed an assay to measure the stability of the HIV capsid. Using this innovative tool, Aiken was able to show that Sundquist's mutations either *increased* or *decreased* capsid stability.[13] Both variations diminished viral infectivity. 'It's sort of the Goldilocks hypothesis, right?' explains Aiken. The capsid, he showed, needs to be 'not too stable, not too unstable'.

'It was very clear that a few single amino acid mutations in the capsid protein basically killed the virus,' recalls Cihlar.

At the time, Cihlar still viewed the capsid as little more than a canister protecting HIV's genes and enzymes, but he could also see that it was a precision-built structure that had to come together in a certain sequence and at certain exact angles. Such intricacy, he felt, might be disrupted by a small molecule drug.

His colleagues were dubious. Drug hunters like targeting enzymes because they have evolved to bind to molecules. Structural proteins like the capsid protein are generally thought not to have the same druggable binding sites.

Moreover, the high concentration of capsid protein that accumulates in infectious HIV particles presented a challenge: a well-designed small molecule might take out some of it, but most researchers thought that reaching all of it would likely require intolerable concentrations of the drug in the patient's bloodstream. One molecule discovered in 2003 seemed to bind to the capsid, and was claimed to have weak antiviral activity,[14] but that line of enquiry had led nowhere. Most of the HIV research community considered the capsid undruggable.

'There were many sceptics that said, "Well, this will never work,"' recalls Cihlar.

Nevertheless, he persisted, reading all the literature on the HIV capsid, visiting Sundquist's lab in Salt Lake City, and slowly building the case for a capsid-targeting programme. The two men became friends, and they both agreed that if Gilead could find a small molecule that interfered with the assembly of the capsid, the company might have a shot at creating an entirely new class of HIV antivirals.

The curiosity instinct runs deep in Stephen Yant. As a child, he was fascinated by the natural phenomena he read about: animals that hibernate all winter, starfish and axolotls that can regrow whole arms. He's always loved discovering stuff. A biology degree led to work on transposons, sequences of DNA that can seemingly 'jump' from one part of the genome to another. During his postdoc at Stanford he became interested in viruses, but until he interviewed at nearby Gilead, he'd never worked on HIV.

'He stood out among all the candidates for his kind demeanour and genuine curiosity,' recalls Cihlar. At the end of a long day of interviews, Cihlar drew a picture of the HIV capsid on a white board. This, he told Yant, would be his target if he joined the company.

Gilead had a library of around 450,000 chemical compounds that Cihlar wanted Yant to test against the capsid. But to do that, Yant would need to create a novel assay that could be used in a high-throughput screen.

It was a formidable assignment. How do you create a test that will efficiently and reliably reveal the effects a vast array of different chemicals might have on a microscopic structure? Yant spent two years reviewing different types of assays. He decided to focus on the speed of assembly of the capsid from its constituent CA protein monomers. A compound that slowed assembly – an inhibitor – might be an effective way to disrupt HIV replication.

Drawing upon an earlier discovery that CA protein monomers will self-assemble into tubes in the right, heavily saline, conditions,[15] Yant developed a process for driving capsid tube assembly over a period of a few hours. The tubes diffract light, so the more tubes that form the more light is absorbed. By measuring the level of light absorbance, he could gauge the degree of tube assembly. Then he could add different chemicals to the solution and see if any slowed down tube assembly.

Do that on a grand scale, with robotic pipettes, 384-well plates and automated light readers, and it should be possible to screen thousands of compounds. Yant had never worked with robotics before, and he only had one team member to help run the high-throughput screen, but he was ready for the challenge.

To test so many compounds, Yant needed a large amount of CA protein – fifty grams, to be exact. That may not sound a lot (fifty grams of granulated sugar equals four tablespoons), but you can't simply harvest the stuff from viruses. CA protein has to be grown in genetically engineered bacteria and then separated out from the broth of chemicals and cellular particles. Just obtaining a few micrograms was a challenge. Fifty grams was an epic undertaking.

A colleague in Gilead's protein chemistry group, Roman Sakowicz, suggested a possible solution. Previously, Sakowicz had worked on cancer drugs targeting cellular structures called microtubules. Like capsids, these structures could spontaneously self-assemble from small proteins, given the right conditions. Remove those conditions and they disassemble. Sakowicz had previously exploited this property to purify microtubule proteins, and he proposed trying the same trick with CA protein.

The protein chemistry team grew CA protein in bacteria, and then added salt and other chemicals. The CA protein rapidly assembled into tubes, falling out of solution to form a white solid that could be easily pelleted. The team washed away the impurities, leaving just the tubes. Finally, they removed the saline conditions, causing the tubes to disassemble into soluble CA protein monomers. Repeating this process several times – assembling, washing and disassembling capsid tubes – left highly purified CA protein.[16]

'We were able to make use of the natural function of the protein itself to help purify it,' says Yant.

With a clever light-absorbance-based methodology and a batch of pure CA protein, Yant and his colleague were ready to start the high-throughput screen. It was, he admits, 'a little nerve-wracking', because he'd never done anything like it before. They would be under considerable time pressure: these were live assays, so once the tube assembly reaction started they had to be on hand until it was finished. They couldn't press pause and go to lunch or they might miss a critical measurement, or be absent when one of the robotic components inevitably failed. Their colleagues would come by to offer food and encouragement from time to time, and that was enough support to see them through. Yant remembers it as an exhilarating time.

The focus initially was on compounds that might *inhibit* – or slow down – capsid assembly. But Gilead's head of structural chemistry, Swami Swaminathan, cleverly suggested they should also search for compounds that might have the opposite effect and *accelerate* assembly. Swaminathan's logic was that if you rush a building project, you're likely to get a different – probably inferior – outcome. If the capsid assembly was accelerated, perhaps it might be compromised.

So Yant measured light absorbance – the proxy for tube assembly – at two time points, early and late in the process. The late point measurement would indicate inhibition of assembly if light absorbance was lower than usual. The early point measurement would reveal an acceleration of assembly if light absorbance was unexpectedly high.

The extra measurements paid off. Yant's high-throughput screen identified a variety of chemical compounds that seemed to impact capsid assembly. Most of them were *inhibitors*, retarding assembly. But a few seemed to be *accelerators*, speeding up assembly. When Yant tested these inhibitors and accelerators on HIV in a cell culture, both groups interfered with viral replication.

The team strengthened the evidence that these compounds were acting on the capsid by following the standard approach of studying strains of HIV selected *in vitro* for resistance to several of them. When they sequenced the genomes of the resistant HIV strains, the genetic mutations were all in the region of RNA that encoded the capsid. By showing what the virus needed to change in order to escape the impact of these compounds, they demonstrated the direct connection between their molecules and the capsid.

Tomas Cihlar was delighted: this was evidence enough to commit further Gilead resources to the project. The antiviral potency of the compounds identified in the high-throughput screen was relatively weak, and most were unstable or had poor PK properties. Cihlar now hoped the medicinal chemists could work their magic, discovering molecules with increased potency and improved oral bioavailability and metabolic stability.

At an early age, John Link found himself frustrated in biology class. He liked the subject, but whenever teachers would drill down into a biological process and explain what was really happening inside a body, they would stop at the molecular level.

'They would say, "And then there are these molecules, and they are at the core of the process, but this is biology so we won't talk about the molecules." I thought, that's where all the action is happening – I want to learn about molecules!'

Biochemistry seemed to be the answer, he decided, if he wanted to get to grips with those all-important molecules. But when he consulted a biochemist at college in Minnesota, he was told, 'You really need chemistry first, as an underpinning.'

He never quite made it to biochemistry class. Pure chemistry was pure joy to Link, and once he'd mastered the discipline, he found he could read and understand biochemistry texts just fine on his own.

Link's interest in biology hadn't waned, so after a PhD at Harvard, he chose a career in drug discovery. The invitation to join Gilead's capsid-targeting programme as the first chemist answered a deeply personal need. A close family member had died of AIDS just before the advent of HAART. 'My cousin was a wonderful, open, really great person. I liked him a lot,' says Link. 'My wife and I would visit him in San Francisco and see him with his friends. I really saw a lot of the adversity people had to face – the stigma and discrimination.'

His cousin's partner also had AIDS, but had survived long enough to get on the HAART regimen, saving his life. The difference in outcomes for the two men hammered home the importance of good HIV drugs for Link.

Despite the new antivirals, Link knew his cousin's partner was in a precarious position. The virus in his lymphocytes had already developed drug resistance. 'I was having dinner with him and he said to me, "You know, I have one mechanism left in my back pocket."' Link understood exactly what that meant: if any of the medicines he was currently on stopped working, there was only one type of antiviral standing between him and likely death.

Link vowed to give him another.

He had no illusions about the task ahead. The capsid was undoubtedly the most difficult HIV target to hit. 'I thought to myself, *Wow, this is incredibly daunting*. I mean, at that time there was no approved drug that targeted protein–protein interactions.'

Going after the capsid meant finding a small molecule that could interfere with the way one capsid protein monomer bound to another. Such protein–protein interactions were only just coming on to the drug hunter's radar.[17] A small molecule is literally *very small*; it would have to somehow come between two massive protein molecules and interfere with their interaction. 'You're talking about bowling balls and a ping-pong ball,' says Link. 'How do you *do that*?'

It was hard enough to interfere with two protein molecules coming together, but each capsid contains 1,500 capsid protein monomers. It boggled the mind to think how concentrated a drug dose might need to be to hit all those monomers in all those capsids throughout a patient's body. 'We were trying to climb Mount Everest when no one had ever climbed a mountain before. This was the protein–protein interaction target of all protein–protein interaction targets,' says Link excitedly. 'I'm sitting there going, *Wow, OK, this is super-cool. This is really, really cool*.'

Link explains his job as akin to molecular architecture.* Like an architect, he will look at the structure of a molecule and envision what he could change to improve it. 'It's not widely appreciated how much creativity there is in what we do.' Yant's molecules had very low potency, so the first task was to increase their antiviral potential. The chemists needed to strengthen the binding between molecule and target by shaping the molecule to fit better in a CA protein binding site and building in interactions such as hydrogen bonds. It's a slow, incremental process.

'The vast majority of the time, the molecule doesn't do what you want it to do but you learn something from it,' he says.

At first, Gilead's chemists didn't even know if there *was* a binding site on the capsid protein. That changed when they were able to obtain co-crystal structures for several of their inhibitor molecules bound to capsid protein monomers. Nothing like it had been published before: it was the first time they could see in detail how a small molecule could bind to the capsid. Because of this crucial structural information, the team chose to focus on the inhibitor series rather than the accelerators.

However, the binding site that the inhibitors were exploiting was small and lipophilic (greasy) – not at all ideal. It offered the potential

* Jewellery, Lego, architecture – the variety of analogies we've encountered in medicinal chemistry underlines both the complexity of the subject and the creativity of its practitioners.

for just two hydrogen-bonding interactions, not enough to achieve a strong binding. This was not going to be straightforward.

After nearly two years' work, Link and his team had adapted the molecules to fill the little binding site. But they needed to increase potency still further, which meant trying to get a better grip on the capsid. 'What we started doing was making the molecules bigger and just spreading them out on to the surface of the protein,' explains Link. Picture a rock climber reaching for a handhold. There's a tiny indentation in smooth rock: the climber can get the tips of two fingers inside, but it's wet and slippery. Beyond that, all they can do is spread their palm flat across the surface of the rock and hope it provides some slight extra grip.

Not exactly optimal in a life or death situation.

By 2009, Link had gone about as far as he could go with the inhibitor series. Whatever modifications the chemists made, potency barely rose above a minimal level. 'The team was on the verge of throwing in the towel,' says Cihlar. 'We were scraping the bottom of the barrel for ideas.'

Cihlar had continued to research the biology of the HIV capsid during the first four years of the programme. He was more convinced than ever that it could be a viable target. But it was becoming clear to him that they just didn't have the right compounds to do it. 'There was no Plan B, no other option we could interrogate.'

Link politely disagrees. He never considered throwing in the towel. He had a real target and twenty scientists determined to crack it. Reluctantly, he acknowledged that the inhibitor series was a dead end, but that just meant it was time to turn to the accelerators. Perhaps they might point to a better binding site?

As we've seen, most of the CA protein that makes up the capsid is organised in hexamers, rings comprising six monomers. The Gilead team had developed a method to purify hexamers and test molecules against these, rather than the simpler monomers of the initial phase of work. Using a very cool technique called surface plasmon resonance, they discovered that, while their inhibitors bound to individual monomers, the accelerators bound much more strongly to complete hexamers.

Groundbreaking work by Owen Pornillos at the Scripps Institute had yielded the first crystal structure of a CA hexamer,[18] and this allowed the Gilead team to hypothesise how accelerators might act on the capsid. The accelerator binding site seemed to lie at the interface between two monomers within the hexamer. Accelerators, they surmised, clamp the monomers together more tightly, speeding capsid assembly. The result is a somewhat rushed, chaotic process, leading to a disorganised assembly that produces broken capsids. Even if the capsid does form correctly, that clamping makes it more rigid and brittle than is optimal for the virus.

But the team was running out of time. Gilead had meanwhile launched a single daily tablet that combined three HIV enzyme inhibitors: TDF, emtricitabine and efavirenz.* The capsid programme looked increasingly irrelevant and hopeless to some of its leaders. Gilead was on the brink of pulling the plug.

Then, in January 2010, new information helped redirect the team.

Gilead had sponsored the Keystone Symposia on HIV Biology and Pathogenesis in Santa Fe, New Mexico, and Cihlar went along to represent the company. One of the ways scientists communicate new research at such conferences is on a poster, intricately populated with data, graphs and statistics. The posters are affixed to boards in a large hall, and delegates can wander between them and chat with the presenters to explore what's going on in the field.

Strolling among the posters one evening, Cihlar came across one created by Christopher Aiken, the capsid researcher from Vanderbilt. Aiken was no longer in the hall, but Cihlar was immediately drawn to his poster. It described a small molecule, PF74, discovered at Pfizer. It was a capsid-targeting molecule.

Pfizer no longer had any commercial interest in PF74. The company had experienced persistent setbacks during its development and had terminated the project. As we saw in Chapter 6,

* With Bristol-Myers Squibb and Merck.

Pfizer's Antiviral Discovery Group had been disbanded. Several of the researchers who had discovered PF74 had left the company, including the former project leader, Wade Blair.

Cihlar was doubtful the molecule could make a decent drug: it wasn't potent enough, and because it looked like a peptide it would be very unstable in the blood. Still, it was a new chemical class. Cihlar scribbled down the molecular structure and, on his return to California, went straight to see John Link.

'Can you do something with it?' he asked.

Wade Blair originally planned to become a physician, but got sidetracked by the lure of microbiology. 'There's just something fascinating about microbes,' he says. The devastating impact certain diseases have had on humanity and our culture throughout history captivated him. 'I was attracted to the possibility of doing something to stop dangerous pathogens,' he explains.

By the time he began his PhD at UC Irvine, HIV was the absolute focus of every virology programme. This new, deadly disease was spreading so fast and killing so many it was already on a par with the historical plagues Blair had studied as an undergraduate. 'It really sank in when I went to grad school,' he recalls. 'A student in my programme contracted HIV and later passed away. That was pretty devastating.'

Blair was hired into what became the Pfizer Antiviral Discovery Group in La Jolla in 2000.* His mission was to search for novel targets for HIV drugs. The group had previously focused on specific HIV enzymes, but Blair took a different approach, instructing his team, 'Let's let the virus tell us what the new targets are.'

To do this, they would need to screen over a million chemical compounds against the live virus. Any work involving dangerous viruses should take place in a secure lab.† To screen a million compounds, sophisticated robotics are needed to load all the plates

* At the time, the La Jolla site was owned by Agouron, which had recently been acquired by Warner-Lambert, which in turn was acquired by Pfizer in 2000.

† BioSafety Level 2 or above.

and measure the results in thousands upon thousands of wells. The La Jolla site had both a secure lab and robotics – just not in the same place. So Blair's team had to recreate a portion of the robotic screening facility inside the secure lab.

Once they had excluded non-specific inhibitors and known HIV drug families, only one molecule in the entire Pfizer library showed antiviral activity against HIV. The activity was very weak, so Blair and his team began a search for closely related molecules that might have greater potency. The most promising was PF74.

Blair set up a series of standard resistance studies, similar to those the Gilead team would later conduct, in which the team gradually exposed wildtype HIV to tiny, increasing doses of PF74 until the virus developed resistance against the molecule. When they sequenced the viral genome, they found – to their great surprise – that the resistance mutation had arisen in the genes that encode the capsid.

'We were pretty excited,' says Blair. 'It was completely new.' Unlike Cihlar and Yant, Blair had not been looking for a capsid-targeting molecule; but that was what his whole-virus screen had spat out.

Blair reached out to Pfizer's structural biology group, who were able to create a co-crystal structure of PF74 bound to a portion of the HIV capsid monomer. It suggested a new binding site at the interface between two capsid monomers within a hexamer.

So, by 2005 – a year before Cihlar kicked off the capsid-targeting programme at Gilead – Blair had an antiviral molecule that bound directly – if weakly – to the HIV capsid, and he had an idea roughly how and where it bound. Now it was up to Pfizer's medicinal chemists in Sandwich to refine the structure of PF74 to make it more potent.

Blair remained in touch with them, but he was effectively off the project. Moved to other HIV and hepatitis C programmes in La Jolla, he could only watch from a distance as the Sandwich chemists ran into challenge after challenge trying to turn PF74 into a potent oral antiviral. They were not successful in the time frame they were given and, eventually, the capsid-targeting programme was terminated.

It must have been so painful for Blair and his team to see their groundbreaking work canned. Worse was to come as we know: the La Jolla antiviral team was disbanded in 2007. Pfizer would not return to antiviral research until Covid-19 struck more than a decade later.

There was one silver lining to Pfizer's exit from the antiviral field: Blair et al. were finally free to tell the world all about PF74.[19]

PF74 was a useful seed to inspire the chemists at Gilead: it was an accelerator from an entirely new chemical series. But John Link and his team knew they would have to create a very different molecule if they wanted to make a viable medicine. The challenges they faced were legion. First and foremost, they had to increase potency so the drug would have significant antiviral effect at a manageable concentration in the blood. There were toxicology challenges too. PF74 damaged the CYP enzymes responsible for oxidation in the liver, potentially leading to harmful accumulations of other drugs a patient might be taking. Such drug–drug interactions are problematic at the best of times, and likely to be disastrous in an AIDS patient on multiple drugs.

But the biggest problem of all was stability. PF74 was rapidly cleaved in the bloodstream and also swiftly oxidised in the liver. 'It was fully metabolised within three minutes in liver cells,' says Link. 'This thing was the most metabolically unstable molecule I had ever seen.' Pharmacokineticists measure drug stability in various ways, but one key metric is *hepatic extraction* – how much of the drug is eliminated on one pass through the liver. For PF74, it was 100 per cent.

Link likens the stability problem to pumping water through a leaky pipe: if you have enough leaks, nothing will come out of the other end. For PF74, there were so many natural mechanisms tearing it apart, it would last barely a few minutes in a human body. None would reach the end of the pipe – the viral target.

'Medicinal chemists within the department were saying, "We shouldn't work on this, it's just not a viable molecule,"' recalls Link.

But he was determined to try. Spurred on by the memory of his cousin, Link would not rest in his effort to discover a new HIV antiviral mechanism. People were still dying, and the longer it took to find a successful molecule the more lives would be lost. He felt he had to go all in. 'There's a moment in life when you realise it's *your moment*. It's your moment to make a difference. And then you do everything you can.'

The chemists were guided by new co-crystal structures. At first the images included only one monomer, showing half of the capsid binding site. Later, they were able to obtain co-crystal structures with a full hexamer, revealing the entire binding site at the interface between two monomers. They could see all the binding interactions between the capsid and an accelerator – another scientific first.

Nonetheless, the capsid project remained on shaky ground and Gilead was considering other alternatives. Late one night at a virology conference in Bulgaria, Link was having a drink with Cihlar:

'Tomas, I want to make a pact with you,' he said. 'If we get wind of plans to kill the programme, we need to walk into the head of research's office and threaten to quit. It's the only way to keep the project alive.'

Immediately, Cihlar shot back, 'John, you know all that will accomplish is that we both get fired.'

They smiled at each other. Downing their drinks, they agreed to the pact.

Over the course of six long years, Link's team of chemists laboured to solve all the various problems they'd identified. Drug discovery takes a long time, as we know, but it's almost unheard of for the medicinal chemistry alone to take six years – on top of the years spent previously on the inhibitor series.

It was a gruelling time for Link, who was simultaneously leading projects in Gilead's hepatitis C programme. He was working literally every waking hour of the week. 'I talked to my wife at the beginning and said to her, "I may not see you for a year. Are you OK with that?" She said, "It's important, you should do it."' In the end, it was worse than he'd predicted. 'I pretty much didn't see my wife for three years.'

In all, the chemists designed and hand-made over three thousand different compounds. At times, it felt like a game of Whac-A-Mole: they would improve potency with one bit of molecular surgery, but that would make the instability worse. They would add a group to the molecule to protect it from oxidation, but that would interfere with binding, causing potency to plummet. It was deeply frustrating.

Continuing with Link's leaky pipe metaphor, one of the difficulties was knowing whether they were actually succeeding in patching any of the holes. They would make a tweak to improve stability, yet still no water would come out of the end of the pipe. Did that mean they hadn't patched a hole? Or were the other holes still numerous enough to empty all the water? Or had they in fact made a new hole? With all of their test compound being wiped out so quickly in metabolic assays, it was hard to know if they were making any progress.

But they stuck with it, adding an atom here, a molecular group there, until after six years they had a molecule that was twelve thousand times more potent than PF74 and had a hepatic extraction of less than one per cent. X-ray crystallography showed it fit in the same binding site as PF74, but the molecule was much larger and vastly more complex. It filled every nook and cranny of the binding site, and with seven hydrogen bonds its grip on the capsid protein was much more powerful.[20] The resulting super-potency meant that even minuscule amounts of the drug in the bloodstream should be enough to stop most HIV replication. The half-life of the molecule in animal models was measured in days, not seconds as for PF74. It was incredibly stable.

'We got it down to be so metabolically stable that it was not measurable,' says Link, referring to the standard metabolic assays.

In the end, the team tagged the molecule with a radioactive marker to achieve a finer level of measurement and demonstrate the almost negligible level of hepatic extraction. When injected under the skin of rats, it was found to last for *months* without degradation.

The finished molecule was labelled GS-6207. It is a network of carbon-based rings studded with fluorine, sulphur and chlorine atoms. To medicinal chemists, it looks enormous, and profoundly odd. Link smiles as he remarks, 'I've had chemists come up to me and say, "You worked on GS-6207 . . . Where did it *come from*? It looks like an alien molecule." '

Eventually it would be christened lenacapavir.[21]

Six years on, the Gilead team had formed a much clearer picture of how accelerators like lenacapavir work against HIV. We saw earlier that accelerators hold the monomers within a hexamer together more tightly, leading to a more brittle, rigid capsid. This unnatural rigidity seems to make it harder for the capsid to enter the host cell nucleus. The capsid has to pass through a *nuclear pore*, a hole in the nucleus, in order to deliver its genetic payload to the host genome.

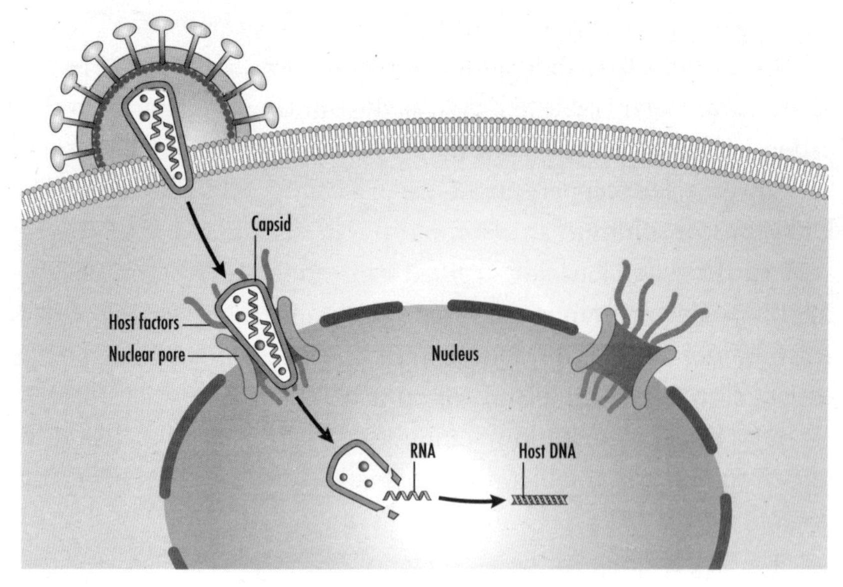

Figure 24. Entry of the HIV viral capsid into a cell nucleus.

'But the capsid is fairly big and the pore is narrow, so the capsid needs to *squish* through this little hole,' explains Yant. Rendered more rigid by an accelerator binding its monomers together, the capsid is less good at *squishing*.

Accelerators frustrate HIV entry into the nucleus in another way, too. We now know that little proteins called *host factors* on the inside and outside of the nuclear pore function like docking arms on a spaceship to lock on to the capsid and tug it through the nuclear pore. And where do these tentacles lock on? The very same binding site that accelerators fit into.

In other words, lenacapavir both rigidifies the capsid and blocks the host factors trying to haul the capsid into the nucleus.[22]

This is why lenacapavir is unique among current HIV antiretroviral drugs: it blocks multiple, distinct stages of the HIV life cycle. Early in the life cycle, lenacapavir prevents capsids from entering the nucleus, blocking the delivery and integration of HIV genes into the host genome. Late in the life cycle, it interferes with capsid assembly, leading to malformed capsids and so blocking the formation of new virus.[23]

Yant and Link compare HIV drugs to different players on a soccer pitch trying to prevent the other team scoring a goal: the goalkeeper, full backs, centre backs and midfielders can all block a goal in different ways. Lenacapavir is able to play multiple roles in blocking HIV replication – it can play goalkeeper, full back *and* defensive midfielder.*

However, lenacapavir posed a significant practical challenge: it had very low solubility in water.

By 2016, when lenacapavir was discovered, the standard of care in HIV was well established, largely due to Gilead's own successes with combinations of antiretrovirals: people with HIV now expected to take one tablet once a day to suppress the virus.

But a molecule with low solubility in water can be challenging to deliver as an oral tablet: if it doesn't dissolve in the stomach, it

* Link is quick to qualify this sporting analogy: lenacapavir isn't doing every job on the team; other drugs are still needed in combination for treatment.

will not be absorbed through the gut. The chances for absorption seemed poor anyway, given its large molecular size. Medicinal chemists are well versed in the 'rules' for orally bioavailable drugs. *Lipinski's rule of five* postulates a maximum weight of the molecule, a maximum surface polarity and a maximum lipophilicity (greasiness), beyond which the odds of achieving an oral drug are low, based on statistics from prior drugs. Lenacapavir breaks all of these rules: it's much bigger, more polar and more greasy than it should be for easy oral absorption.

The molecule also breaks a bunch of other rules established to guide drug hunters towards viable small molecule medicines, but Link considers such dogma rather limiting. 'If you are trying to solve a problem that's never been solved, you can't look backwards,' he says. 'We didn't even think about the rules.'[24]

They had to make the molecule abnormally big, he explains, to have any impression on the capsid's protein–protein interactions (remember the ping-pong ball between two bowling balls), and to defend it – or armour-plate it – against metabolic destruction. Inevitably, that made it less soluble.

But every challenge can be turned into an opportunity for innovation with some creative thinking, and that is exactly what Gilead did. Lenacapavir is incredibly potent and stable, so a very small amount of the drug might be able to sit in a human body for months without degrading and still have a massive antiviral effect. *What if*, the team asked themselves, *we inject it under the skin and just leave it there to dissolve very, very slowly?*

'We were making a compound that was super metabolically stable and ridiculously potent but we were never moving the needle on the solubility – if anything it was getting worse,' says Yant. 'And then we looked into the possibility of making that an asset.'

It was a huge departure from the original goal of a once-daily tablet. By now, people with HIV had a range of pharmaceutical options to manage their infection, but all of them required daily medication. Wouldn't it be a whole lot better to get a painless injection and then not have to think about the disease for months?

That was a truly exciting idea. A drug that might only need to be administered twice a year would be a major improvement in therapeutic experience for the millions of people living with HIV. But could it be done?

Roshy Pakdaman used to teach physical chemistry as a university professor in France. She loved spending time with students but gave up the academic lifestyle in order to make a greater impact in the pharmaceutical industry. Her more than two decades at Gilead have amply paid out on that life choice, resulting in a string of important formulation innovations. Now she's in charge, among other things, of formulating molecules like lenacapavir.

Pakdaman has a charming, surprising laugh, which she deploys regularly as she describes the difficulties of dissolving lenacapavir in water-based liquids. Formulators sometimes refer to such difficult-to-dissolve molecules as 'brick dust'. It was up to Pakdaman and her team to figure out a formulation that could be injected. They had to find a way to make the brick dust more or less liquid.

Initially, rather than try to dissolve lenacapavir, they mixed a small amount of undissolved particles in water with some additives. This would yield only a low effective dose, but it would suffice for the first-in-human study.

In charge of the clinical trials was Martin Rhee, a physician from South Korea who'd moved to the United States to advance his medical training and had ended up staying for decades. He vividly remembers the first patient he treated with HIV. The man had terrible vertigo: for months, he felt like the room was spinning. No one could work out what was wrong with him. Eventually he was diagnosed with HIV; the vertigo was caused by a related brain lesion.

In planning the Phase I trial of lenacapavir in healthy volunteers, the big challenge for Rhee and his team was how to manage the safety risks inherent in a drug that is so stable it might linger in the body for a very long time. 'What if they develop some acute reaction?' he said. 'With an oral daily drug you can immediately stop. Here, you can't stop! You know this drug concentration is going to last for potentially months.'

So Rhee planned a cautious study, administering very low doses to just two healthy volunteers and then waiting a long time to check for adverse reactions before giving the next volunteers a slightly higher dose. Thankfully, lenacapavir seemed to be well tolerated, with only mild reaction at the injection site.

Even better, the drug seemed to last a very long time in the volunteers' bodies. 'Given our hopes for a twice-yearly medicine, we were so excited to detect lenacapavir in the blood six months after a single injection,' says Rhee.[25]

Next up was the Phase Ib proof-of-concept study: lenacapavir was administered to a small number of people with HIV to test its antiviral activity in a human body. The early results were promising. Lenacapavir caused a 100-fold decrease in viral load. Very few HIV drugs ever achieve that level of potency on their own. Moreover, it had a predictable dose-response relationship: the more you administered, the more antiviral effect you got, up to a maximum point. That's a very desirable trait in any drug.

Meanwhile, Pakdaman and her team were working on an alternative formulation that could load more lenacapavir into the small volume of fluid that could be injected under the skin.

The physical and chemical properties of a drug and its *sodium salt* (a derivative of the drug) are often different, and this difference can be exploited for drug delivery purposes. Pakdaman's team were able to develop a reliable formulation that could dissolve large quantities of lenacapavir sodium salt. When this formulation was injected subcutaneously, small particles of lenacapavir would form in the fat under the skin. These particles would then slowly dissolve and disperse throughout the body.

After injecting this formulation in the next study, the clinical team measured the concentration of lenacapavir in the volunteers' blood at regular intervals. As expected, it lasted a long time. In fact, a very long time. Lenacapavir's half-life, they calculated, was a staggering fifty-two days: two months after injection, nearly half of the injected drug was still available to fight the virus. And with such high potency, even small quantities suppressed HIV. Reviewing the clinical data, Cihlar and Rhee could see that their goal was within reach: physicians might only need to inject an individual *once every six months* to protect them against HIV.

For the millions of people infected around the world, that would represent a mind-blowing paradigm shift in treatment options.

Well . . . except for the first few weeks. The one remaining problem, inevitable in a slow-release drug, was the initial time required for enough of it to get into the bloodstream from the subcutaneous deposit. Those first few weeks would be critical: if the concentration of lenacapavir were to build up only slowly, then the virus would be reproducing in the presence of low, slowly increasing drug levels – a perfect environment to select for resistance mutations that could undermine the drug from the outset.

What was needed was a *loading tablet* formulation that would be rapidly absorbed until the injectable kicked in. Pakdaman and the formulation team went back to the drawing board. The answer, they discovered, lay in a technique known as *spray-dried dispersion*. This allowed them to make lenacapavir tablets that would dissolve quickly in the stomach. Despite breaking Lipinski's rules of oral bioavailability, with some clever formulation thinking, orally dosed lenacapavir was able to achieve blood concentrations that could rapidly suppress HIV.[26]

Because lenacapavir has a completely different mode of action to other HIV antivirals, it was potentially extremely valuable as a therapeutic of last resort for heavily 'treatment experienced' people with HIV, where the virus has evolved resistance to most existing drugs. When the FDA granted Breakthrough Therapy Designation in May 2019, they encouraged Gilead to focus on this group first.

It's actually quite hard to locate such people, so a study of hundreds of patients was not possible. Nor was it necessary in the view of the FDA, if the dramatic treatment effect seen in Phase Ib could be achieved again. But the regulators did need to see how lenacapavir performed as part of a combination therapy. Gilead had always known their new drug would not be used as a treatment monotherapy, due to the dangers of resistance emerging. 'It's very risky to use a monotherapy in people who have replicating viruses,' explains Rhee. His Phase II/III study would assess lenacapavir in combination with other HIV antivirals.

It took a year to enrol seventy-two participants at forty-two sites across eleven countries. Participants remained on their existing (if somewhat ineffective) medication; for ten days, some of them also received lenacapavir, while the rest were given a placebo. The FDA had set a minimum threshold of a three-fold decline in viral load to demonstrate antiviral activity. When the results were unblinded, eighty-eight per cent of participants on lenacapavir had achieved this while only seventeen per cent on placebo made it. Most people receiving lenacapavir had achieved far more than the target three-fold decline.[27]

The study continued for another year, although now all participants were given lenacapavir. With that initial proof of efficacy, it would have been unethical to continue to administer a placebo. After six months, there were no serious safety issues, and viral load suppression was significantly better than under the patients' previous regimens. After a year, eighty-three per cent of participants receiving lenacapavir in combination with an optimised background regimen had achieved an undetectable viral load.[28]

Lenacapavir was approved for use in adults with multidrug resistant HIV in the UK and the EU in August 2022,[29] and in the United States in December 2022.[30] Gilead is now testing additional long-acting molecules from different classes of HIV inhibitors to combine with lenacapavir in twice-yearly treatment regimens.

That brings us just about up to date. But as a coda, we can peer a little way into the future for another possible – and very exciting – application of lenacapavir.

When Moupali Das heard about the stability data coming out of the lenacapavir studies, she was blown away. 'I was like, this drug, you can give it every six months, that's crazy! That's basically like an *HIV vaccine*.'

Das has an almost lifelong commitment to addressing the injustices she sees in HIV healthcare provision worldwide. An infectious disease physician, she was highly influenced by the humanitarian doctor Paul Farmer, whom she met as an undergraduate. Farmer co-founded the non-profit Partners in Health, and worked throughout his life to fight disease in some of the most disadvantaged parts of the world. Das likes to reference his insight that infections track along the fault lines of social inequality. So it has been with HIV.

The first injustice she registered, while still a sophomore in high school, was the story of Ryan White, a boy with haemophilia who contracted HIV through a blood transfusion and was subsequently denied entry to his school. 'I remember hearing about him and thinking how unfair it was. We didn't really understand a whole bunch but we knew that HIV wasn't spread through somebody being in your classroom.'

Although she grew up in the United States, most of Das's extended family lived in India, and prior to med school she got a travelling scholarship to study women's health there. In Kolkata, she worked at an HIV prevention clinic run by a union of sex workers. 'That was extremely eye-opening and fascinating,' she says with a smile. 'We had to hide from my grandmother that I was going to the red light district.' Whilst working there in the summer of 1996, she read about the advent of HAART therapies for HIV; the sex workers were pessimistic, predicting, 'Oh, they're never going to come to India.'

While at med school in New York, Das realised you didn't need to go to Africa or India to find similar health injustices. She saw significant disparities in HIV treatment in the Bronx and Brooklyn. 'There were people in our own backyard who weren't benefiting at that time from antiretroviral therapy,' she says fervently.

'Communities of colour, migrant communities, people left out were in despair, and that continues unfortunately to this day.'

One possible answer to these health injustices is to offer a form of prophylaxis to such at-risk communities. Pre-exposure prophylaxis (PrEP) gives HIV drugs to uninfected people to reduce their risk of contracting the virus. Men who have sex with men are doing very well on PrEP. For the first time, the rate of new HIV infection is lower in MSM than in heterosexuals in many countries.[31] The groups with the highest rates of HIV infection are the marginalised groups sometimes excluded from PrEP for reason of inadequate healthcare provision or stigma – migrants, communities of colour and transgender people.

Das left a highly fulfilling career in public health medicine and joined Gilead partly in order to improve access to HIV drugs for those marginalised groups. That's why she was so excited to hear about the long-acting potential of lenacapavir. Oral PrEP requires people who are healthy to take a daily medicine, reminding them every day of the stigma associated with their risk of contracting HIV. Possessing a pill bottle linked to HIV may also place a person at risk of social or physical harm from family, friends or partners. If lenacapavir could protect people from contracting HIV, it would be a very convenient prophylactic, fundamentally different from a daily pill – functionally, a chemical vaccine.[32]

We could imagine public health programmes providing lenacapavir free to at-risk communities all over the world. This could be the drug that finally ends the HIV/AIDS pandemic.

But before we get there, Das and Gilead need to demonstrate lenacapavir actually helps prevent HIV infection. Das has designed, and is currently running, a series of clinical trials testing the drug in large groups of at-risk, sexually active people.[33] The trial design is challenging, as it would be unethical to give a proportion of the trial participants a placebo we know would not protect them. Instead, Gilead and the FDA have agreed upon use of an innovative approach based on trials for birth control (which have the same ethical problem). It makes use of a *counter-factual* participant group.

The trial team gives lenacapavir to a group of previously uninfected people for a year and measures the HIV incidence (the number of new cases of HIV arising during that period). They compare that to an estimate of the incidence in a counterfactual group of people not on lenacapavir.* A significantly lower incidence in the treated group would indicate that lenacapavir is effective as pre-exposure prophylaxis against HIV infection.

Ever mindful of the potential for injustice in health outcomes, Das and her partners have focused the clinical trials on those groups that have been disproportionately affected, with study sites in Uganda, South Africa and other high-incidence countries. They've set up community advisory groups to build trust among marginalised people, they're including community advocates who represent the participants who would enrol in the trials, and they've redesigned the way Gilead collects race and gender data. 'The science is better if everyone has a seat at the table,' Das says.

We don't have the results from these studies yet. But if they show a meaningful preventative effect, lenacapavir could be the molecule that finally cracks HIV.

'I think it's really going to help end the HIV epidemic for everyone, everywhere,' says Das. 'This really could be a gamechanger.'

John Link is now retired, prematurely in some people's eyes, but he sees that as his answer to the question of work–life balance. After years of sixteen-hour days discovering treatments for HIV and cures for hepatitis C, he is now focusing on the life side of the equation. He enjoys travelling with his wife and playing the guitar. A long-time Hendrix fan, he recently staged a multimedia show for his friends to mark the rock icon's eightieth birthday.

Tomas Cihlar and Stephen Yant continue to pursue new treatments for HIV and other viral diseases, including Covid-19 and future pandemic threats. The pace of drug development is

* They use a recency assay to figure out how recently HIV-positive members of the screening population have been infected, and from that are able to work out the background incidence.

accelerating, and following the success of lenacapavir they're confident of more breakthroughs in the next few years.

'For me, lenacapavir stands out as one of the strongest examples of deep commitment to medical innovation,' says Cihlar. 'We continue to explore its full potential in both treatment and prevention, and it will no doubt bring us closer to ending more than four decades of HIV epidemic. We've only mentioned a few names, but more than a thousand people contributed to the project over almost two decades, as well as the many clinical study participants. They are *all* the heroes of this story.'

CHAPTER 9

How to Innovate

Insights from the Front Line of Drug Development

THE INNOVATIVE MEDICINES IN THIS BOOK alleviate pain, control bleeding, shrink tumours, kill viruses and improve motor function in babies with disabilities. What have we learned from these innovation journeys? Here are my ten main takeaways. The first eight build upon direct lessons from the discovery stories. The last two are derived from my own experiences.

Although we've focused on a few glorious triumphs, the overall success rate of clinical trials is currently less than 10%. And that's for molecules that have made it into the clinic; it does not even count the number of molecules that died in Discovery. It would be remiss of me not to try to help future generations of drug hunters to improve on those low odds. I'm also hopeful that these insights may catalyse new ideas in other fields of study and enquiry.

The insights are solely my own and may not reflect the views of the people who have contributed to this book or any of the companies that agreed to share their stories.

1. Successful innovation in drug discovery happens when we reach the holy grail: the 'sweet spot'

Up to the end of 2022, there had been only about 1,800 new molecular entities (medicines) approved by the US Food and Drug Administration and the UK Medicines and Healthcare products Regulatory Agency (MHRA).[1,2] Virtually every drug ever developed has reached the sweet spot of discovery. That sweet spot lies at the intersection between three key dimensions:

1) our biological understanding of the drug target;
2) our clinical understanding of the disease in relation to the target;
3) and the technological advancement that enables drugging of the target.

Think of this sweet spot as the centre of a Venn diagram.

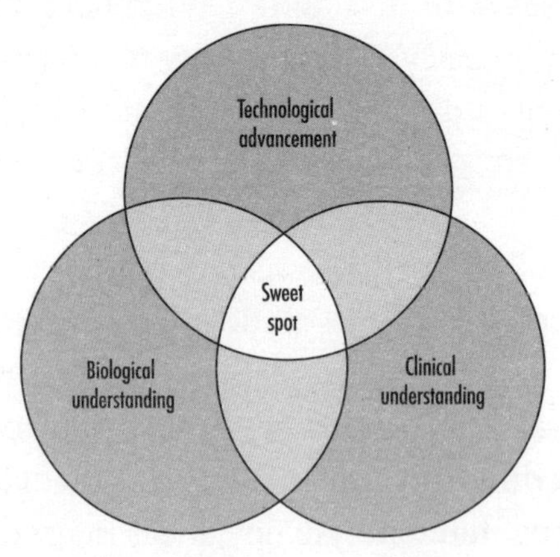

Figure 25. The sweet spot of drug discovery.

Let's use osimertinib as an example. What was known when the molecule started development? From a *biological standpoint*, the target EGFR is a kinase that normally switches on and off to tell a cell when to divide. Mutations in the *EGFR* gene lock EGFR in an 'on' position, leading to continuous cell signalling and uncontrolled cell proliferation. Inhibition of mutant EGFR by a kinase inhibitor shuts off the signalling, inducing cancer cell death. From a *clinical standpoint*, *EGFR* mutations are found in a well-characterised subset of patients with lung cancer who benefit from treatment with EGFR kinase inhibitors. However, over time, their tumours develop resistance to treatment, often through acquisition of another *EGFR* mutation. From a *technological standpoint*, advances in kinase inhibitor chemistry starting in the 1990s enabled rapid design of a safe and effective EGFR double-mutant-specific kinase inhibitor that targets both drug-sensitive and drug-resistant mutations. Putting it all together, *voilà* – we got osimertinib!

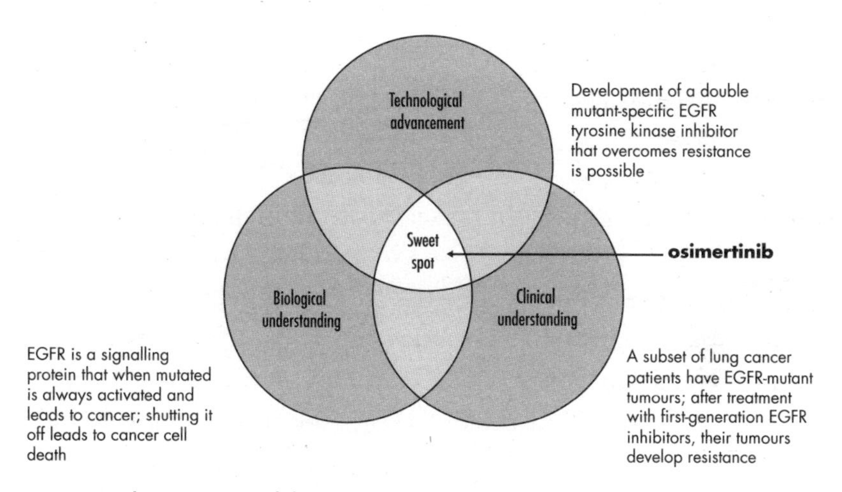

Figure 26. The sweet spot of drug discovery using osimertinib as an example.

It sounds straightforward, but the accumulation of relevant knowledge in all three dimensions usually takes decades to reach a point where drug hunters can even get started. Moreover, it involves multiple experts working in unrelated fields, sharing new-found information with each other at different points of the innovation journey.

We can trace the roots of osimertinib's development back to research conducted by biochemists Rita Levi-Montalcini and Stanley Cohen in the 1950s. Paediatrician Janet Watson hypothesised a potential way to cure sickle cell disease in 1948, based upon her insight into foetal haemoglobin switching. But the idea took another seventy years to become a practical treatment, unlocked by the discoveries of the key role of *BCL11A*, gene editing techniques and, ultimately, CTX001. Similarly, spinal muscular atrophy was first described in the 1890s, but it took geneticist Judith Melki's discovery of the *SMN1* and *SMN2* genes one hundred years later to lay the groundwork for the development of risdiplam and the other two SMA treatments in the 2010s.

If knowledge in one sector of the Venn diagram is poorly developed, it might be wise to wait for it to mature before embarking on a fully-fledged drug discovery programme. For example, if biological understanding of a target is lacking, a drug hunter may choose to hold back until important knowledge gaps are filled. Because of the complexity of the biology, many years may pass before a solid answer arises.

Alternatively, if one has a great biological target it may nevertheless be 'undruggable' because the technology is not mature enough. A drug hunter may need to step back until the relevant technological breakthroughs are made. They should not attempt to apply another, less suitable technology against the target as a matter of convenience (they have the technology in-house and want to use it) or to satisfy a corporate strategy (the company just bought a new technology platform and wants to make the most of it). The particular technology used against a target should be chosen exactly for that target.

Finally, a drug hunter could have a great biological target that is safely druggable, but if the clinical patient population for which the potential drug will be developed is unclear, it may not be sensible to proceed. A 'great molecule in search of a disease' is usually a recipe for failure. It may take a long time to get to the sweet spot – if one gets there at all.

The art of drug hunting – and any activity that requires the application of new technologies to address a complex challenge in the physical world – is knowing when one is close enough to the sweet spot to justify devoting considerable resources to pursuing that innovation. For any potential drug development project, it is important to assess how close it is to the sweet spot and to exercise discipline to steer all activities towards getting to that holy grail.

2. Innovation precedes understanding

Although modern drug discovery aspires to follow a logical flow from scientific understanding to rational drug design to clinical trials to launch, the truth is, despite our best efforts, we often do not have all the answers when we embark on a discovery journey. We often are not yet at the 'sweet spot', with perfect biological and clinical understanding and the right technological advancement. We have to make progress somewhat in the dark, and that progress is rarely linear.

Early drug hunters took an empirical approach to the human body. They tested various compounds against different disorders and, by trial and error, discovered what worked and what did not. The consequences were sometimes lethal. We know far more now, and the process is a lot safer, but as you have seen, drug discovery often still takes a long and winding path to reach a breakthrough.

Over the course of nearly twenty years, Novartis developed multiple PI3K inhibitors for cancer. Various unanticipated clinical issues stymied the first three, and it was only the fourth, alpelisib, that made it to the finish line. Along the way, new insights into PI3K biology (the discovery of *PIK3CA* mutations and of activated PI3K signalling in hormone-resistant breast cancer) enabled a more refined clinical development plan in a specific patient population.

It should not be a surprise that reaching the sweet spot is challenging. 'Biology is not physics,' my PhD supervisor, Adrian Hayday, told me. Equations can explain experimental physics with remarkable accuracy. Force equals mass times acceleration. Distance equals rate times time. Energy equals mass times the speed of light squared. By contrast, it is very hard to foresee exactly how biological systems will react to novel pharmaceutical compounds. There are always unexpected outcomes. That's because we have no truly predictive models that can replicate in a petri dish or an animal what might happen in a human body when a new molecule is administered.

Human beings have evolved through natural selection over millions of years. Each of our bodies has 30–40 trillion cells. Each cell contains a plethora of molecules, signalling pathways, structures and biological processes, many of which we still do not fully understand. On top of that, more than eight billion people live on our planet. Each person has a unique genome, and all these biological components may be slightly different in each person.

As drug hunter Art Levinson, former CEO of Genentech, once told me, 'Drug discovery is not chess, where the list of all possibilities is known.' Chess is one of the most complex games in the world, with an estimated 10^{40} moves possible, at least. Yet we know all the pieces, we know exactly what each piece can do, and there's a finite chessboard. In drug discovery, we don't know the exact characteristics of the 'chessboard', we don't know exactly what each molecule does in each cell, and we don't move just one piece at a time. Drug discovery is infinitely more complex than chess.

3. Innovation demands repeated iteration, refinement and emergent learning

The popular understanding of scientific discovery invokes a few dramatic images that capture our imagination: Archimedes leaping from the bath crying, 'Eureka!'; Isaac Newton watching a falling apple; James Watson and Francis Crick pointing to their double-helix model of DNA. But even when these tales might have a grounding in truth, so-called eureka moments by one or two scientists are the exception, not the rule. More importantly, the moments rarely by themselves deliver a practical innovation. They serve as just one step in a long journey.

To treat sickle cell disease, scientist after scientist went looking for the mechanism that controls foetal haemoglobin switching. They did not know what it was and did not have any idea where to look. Along the way, they searched for an answer in adrenal glands, pituitary glands, transcription factors and other blind alleys. Hundreds of people toiled for nearly six decades before the critical 'Achilles heel' genetic sequence was identified.

Kunihiro Hattori had a eureka moment when he realised that a bispecific antibody against Factors IXa and X might be used to treat patients with haemophilia A. But his team had to synthesise and screen over 44,000 different molecules before they were able to create emicizumab.

Because of the unpredictability inherent in drug discovery, and the need for iterative trial and error, a specific type of learning mindset – emergent learning – should be adopted. Emergent learning makes use of a process of self-guided discovery through iterative experiments: we first identify what we think we do not know, and then we pose specific questions and set up experiment after experiment to deepen our understanding. Along the way, we run into dead ends, answer some questions and generate new ones – hopefully as fast as possible.

Scientists and physicians are often trained in undergraduate, graduate or medical school classes through teacher-centred didactic learning. However, once they start working in a real lab or clinic, they may find themselves in complex new areas, facing time constraints and pressures to present solutions that are not readily obvious or predictable. The dynamic scientific environment requires them to learn from as many sources as possible, constantly, opportunistically and speedily, and to act on those learnings in continuous iterations.

Stephen Yant had never run a high-throughput screen when Tomas Cihlar asked him to test 450,000 compounds against the HIV capsid. He did not know how to grow enough capsid protein, what diagnostic to use to detect efficacy, or how to set up the robotics for the screen, but he worked through all these challenges, built up his knowledge, and successfully set Gilead on the path to creating lenacapavir.

Every project team in this book faced obstacles which they tackled through emergent learning. This highlights the importance in our daily work of learning from each and every experience, whether positive or negative. All experiments should be designed to generate clear answers. Positive results are rewarding, but negative results can be equally valuable as they help us understand why specific approaches do not work and steer us towards better alternatives.

Learning should be shared across teams and organisations, perhaps through dedicated sessions outside of normal decision-making meetings. Such sessions should discuss lessons learned both from drug development projects that were stopped and also from those that made it to launch.

4. Innovation is accelerated by recognising and embracing serendipity

Biochemist Stanley Cohen discovered epidermal growth factor by chance – but his work on nerve growth factor prepared him to recognise what was in front of him. 'Many new things are found by accident,' he once said. 'If you're prepared to see the accident, you can find it.'

The discovery of paracetamol happened purely by chance. Paul Hepp and Arnold Cahn were attempting to treat intestinal worms and prescribed naphthalene for their patient. But when a pharmacist dispensed the wrong compound, acetanilide, and their patient's fever was reduced, Hepp and Cahn were alert to the potential of their accidental discovery. They took immediate steps to investigate the molecule's antipyretic properties. Their discovery of acetanilide's therapeutic benefits eventually led to the development of paracetamol as a medicine.

Similarly, it was fortuitous that nirmatrelvir contained a trifluoroacetamide group that could be used to track the drug through the body without traditional radioactive tagging. Amit Kalgutkar and his Pfizer colleagues realised that this unusual property could be exploited to shave as much as a year off the development time of the much-needed Covid-19 antiviral.

Serendipity, by its nature, is very hard to plan for. Innovators need to be ready to recognise chances that may arise at any time, from anywhere. One way they can do this is by staying up to date on cutting-edge knowledge in their field so they can better recognise fortuitous results and act accordingly. Innovative organisations can give serendipity a helping hand by hiring and developing people who are curious and open to new ideas. Curiosity can also be enabled by carving out time for scientists to explore ideas and to connect with others across different disciplines.

5. Innovation may require challenging accepted orthodoxy

Chemist Hasane Ratni broke a number of rules for the design of an orally bioavailable drug in the development of risdiplam. His colleague, the toxicologist Lutz Müller, believes that scientists who are not willing to go off-piste once in a while are less likely to achieve big scientific breakthroughs. Chemist John Link completely ignored the famous Lipinski rules, among others, when he made lenacapavir. In his view, rules are a brake on creativity.

Rules that become scientific orthodoxy can stifle innovation. Paracetamol was first synthesised in 1878, but wasn't widely used as an analgesic until the 1950s because a prestigious scientist, Joseph von Mering, issued an erroneous toxicology report on the molecule. Von Mering was so respected that no one questioned his finding for decades. Only after Julius Axelrod and Bernard Brodie, among others, performed their landmark research into paracetamol in the late 1940s did the flawed dogma collapse, paving the way for McNeil Laboratories to create a new painkiller.

Breaking with scientific dogma is very hard. It takes breadth and depth of knowledge. It means knowing enough to have the confidence to challenge the status quo. It also takes enormous courage, as professional reputations can be at risk. When Lew Cantley announced the discovery of PI3K, an enzyme his peers said couldn't exist, he was not believed or given funding for years.

In each of these cases, accepted scientific dogma was finally overturned by new information. The best way for a scientist to challenge existing paradigms is to generate new and convincing data.

This approach is codified in the motto of the Royal Society, the UK's scientific academy: *Nullius in verba*, or 'Take nobody's word for it.' The Society's website explains that the motto is 'an expression of the determination of Fellows to withstand the domination of authority and to verify all statements by an appeal to facts determined by experiment'.

6. Innovation means turning an idea into something of practical value on a large scale

A common misconception about scientists and physicians who make breakthrough discoveries in the laboratory or in a handful of patients is that their work is then complete. Of course, their discoveries are crucial, but proving a therapeutic concept is just the beginning. To turn that prototype into a viable drug that can be manufactured at scale, we need to create a cost-effective industrial process that ensures that every single tablet or vial is safe, contains the right dose of the relevant therapeutic, and can be delivered to patients around the world. Recall David Altshuler who contrasted building a Lego bridge with trying to span the Hudson River.

Kunihiro Hattori and Takehisa Kitazawa were able to make a bispecific antibody in the laboratory that effectively mimicked the function of Factor VIIIa in the coagulation cascade. But emicizumab would never have become a medicine if Tom Igawa and Hiroyuki Tsunoda had not solved the chain association problem, so that they could synthesise enough of it to deliver at scale. Lenacapavir offered a completely new way of targeting HIV. But without the formulation expertise of Roshy Pakdaman, developing both a long-acting injectable and a loading oral tablet, it would not have been a viable choice for millions of people with HIV.

In both examples, the scientists drew on new technological breakthroughs to enable mass production and scalability. To deliver more breakthroughs to patients faster, we need to put in place more incentives that stimulate advances in scalable drug manufacturing and delivery, especially for promising new technologies that are getting close to reaching the sweet spot of drug discovery.

7. Competition accelerates innovation

'There's nothing like competition to get things moving,' says Matthew Meyerson, one of my rivals in the EGFR mutation hunt. And getting things moving in drug development can matter a great deal, as a few months or years shaved off a drug's development timeline can impact thousands – perhaps millions – of lives.

There was a competitive race to invent painkillers in the late nineteenth century, kinase inhibitors at the beginning of this century and SMA therapies in the 2010s. Now we see multiple companies competing to create a gene-based cure for sickle cell disease. Ultimately, this is good for patients. Thanks in part to competition, three differentiated SMA therapies became available for patients in the space of just four years.

Competition is also good for science. Swee-Lay Thein published on the *association* of *BCL11A* with foetal haemoglobin switching, scooping Vijay Sankaran and colleagues, but Sankaran was ready to take the next step and prove *causation*.

My own experience of this phenomenon was painful but motivating. Two other groups found drug-sensitive *EGFR* mutations around the same time as our MSKCC team did, and they published ahead of us. The rivalry pushed us to work even harder to confirm and extend the findings and then to identify a key mechanism of drug resistance, ultimately paving the way for the development of osimertinib.

Emergent learners can thrive in a competitive environment, as they are motivated to refine and iterate faster. The efforts made by AstraZeneca scientists to speed the development of osimertinib in the face of a competitor programme at Clovis Oncology exemplified this phenomenon. The Clovis drug was ultimately shelved, but that programme spurred AstraZeneca to get their own drug to patients two or three years earlier.

However, speed must be balanced with quality. The AstraZeneca team understood this, and when a possible safety issue arose concerning the insulin receptor, Susan Galbraith wisely took the decision to fix the problem at the expense of several months' progress.

8. Discoveries in unrelated fields unlock innovation

Generations of molecular biologists dedicated years of their lives to discovering the genetic secrets of foetal haemoglobin switching, but in the end it was a Sardinian population survey of general traits that revealed the answer and made a treatment for sickle cell disease possible. The haemophilia drug emicizumab would not have been possible but for the invention of bispecific antibodies. Ritonavir, an antiviral developed to combat HIV, made nirmatrelvir viable as a Covid-19 treatment. If a Novartis team in the UK hadn't developed an alpha-selective PI3K inhibitor as a possible treatment for respiratory diseases, the company's Oncology group in Switzerland might never have developed alpelisib for breast cancer.

These stories highlight the importance of cross-disciplinary research and fertilisation. Breakthroughs can come from anywhere, so while deep expertise is important, at least one person on any team should be more of a generalist and keep an eye on what's going on in other fields. We can also keep friends and contacts abreast of our innovation goals, even if their own work seems quite unrelated. These people may just hold the key to our next big idea.

Organisations can do this on a bigger scale by hiring diverse talent and inviting outside experts to bring left-field ideas, discoveries and developments into view. Organisations can also give time and space to employees to gain new knowledge in disparate fields. Leaders should encourage cross-functional communication and productive job rotations. Employees, in turn, need to keep abreast of what is going on outside of their own daily work and be willing and excited to take on ideas from a range of sources.

9. Innovation demands determination, resilience and collaboration

The process of drug discovery usually involves trial and error, ups and downs, iteration after iteration after iteration. Thus, in addition to knowledge and skill, important attributes displayed by innovators include purpose, stamina and resilience. Organisations charged with delivering breakthroughs should hire and nurture people who embody these traits.

Such individuals are not always the easiest employees to manage. Sometimes, they come with strong wills and opinions that can be hard for others to tolerate – difficult personalities, a tendency to deviate from established processes, and so on. These behaviours can be tolerated to some extent, as long as the individuals make data-driven decisions and don't negatively impact others or the culture of innovation around them. The ideal innovator is no prima donna but combines intellectual assurance with a touch of self-deprecation, a willingness to learn from others, and the ability to make their team better.

These innovators are what Safi Bahcall calls 'artists' – the dreamers who pursue the big, high-risk ideas that spark that innovation, as opposed to the 'soldiers' – the people who execute well to deliver innovation. While this inevitably simplifies a complex picture, it is a useful way to look at the different talents found in an innovative organisation. Drug development needs both artists and soldiers, and organisations must be sure to offer the right environment and incentives to attract both types of people.

Although the stories in this book highlight only a few individuals from each project, drug discovery is actually the ultimate team sport. By *team* I mean not only an internal company team but the wider team of experts outside the company as well. Thousands of experts working in different areas need to come together over long periods of time to convert an idea into a medicine. Collaboration is essential. In my opinion, the most highly functioning teams are comprised of experts who say what they mean, are accountable to each other, deliver for each other on time, and pull in the same direction once a decision has been made.

Leaders need to pay close attention to teams within their organisations and ensure that individual talents, matrixed teams and individual departments are incentivised for collaborative team behaviours.

10. We can improve the odds of successful innovation through better decision-making

Over the lifetime of a molecule in development, multiple decisions need to be made daily, equating to thousands of decisions made along the way. Time, resources and potentially patients' safety are at stake. Ultimately, one wants to make the most rigorous scientific, data-driven decisions, in the fastest way possible, to enable better outcomes. We can improve the odds of success in innovation by optimising decision-making through fostering honest debate and eliminating bias.

Emerging scientific data is often hard to interpret due to seemingly conflicting results; one group of scientists may have one interpretation, while another group takes the opposite point of view. Under such circumstances, it is vitally important that everyone conducting drug discovery activities feels safe to speak up and voice their opinion, whether popular or unpopular, no matter their rank or position.

In project team settings, it may be particularly hard to voice a contrary opinion, because individuals on the team may not want to be held responsible by their team members for being the one who 'killed' the project. Similarly, individuals don't like to be blamed for an outcome that did not go well as a result of their function's work. Nevertheless, it's most beneficial for decision-making when all points of view are heard and the team takes collective accountability for their work.

The best discussions occur when they are conducted in an environment of intellectual friction without social friction. That is, discussions should be objective and fact-based while minimising personal agendas or individual grudges.

Another way to enable better decision-making is to eliminate bias. Bias, whether conscious or unconscious, arises in many forms. Team members who have spent years of their lives on the project may become biased towards data that supports their hypothesis while choosing to ignore contradictory findings. Companies may become biased towards their own internal data, discounting information from competitors. Leaders who manage multiple projects and budgets may become biased to cut programmes that are not moving fast enough in order to achieve their portfolio milestones.

A more subtle form of bias may occur when the goals of a project team within a larger company environment deviate from their core scientific goals because of 'competing' biases. That is, the project team's scientific goals become distorted by other seemingly important goals, such as corporate strategy goals, that have no direct connection with the discovery project itself. All of these factors can negatively impact scientific decision-making. While biases cannot be eliminated, all employees should be trained to recognise biases in their various forms and should try to minimise their impact on the science to the fullest extent.

* * *

Summing up, then, for drug hunters: concentrate your efforts on reaching the holy grail – the 'sweet spot' intersection between biological understanding of the target, clinical understanding of the target and technological advancement enabling drugging of the target. Find your winning team of purpose-driven and resilient talents who thrive on emergent learning, are ready to pounce on serendipitous findings and boldly challenge orthodoxy when needed. Finally, optimise good scientific decision-making, daily, by fostering open debate and eliminating bias.

I hope the eight stories in this book will inspire and enable new generations of drug hunters in their quest to make transformative medicines. The stories may also provide insights to anyone seeking to make breakthroughs in other fields. The ideas, tools and techniques that enable innovation are all around us. It's up to us to go and find them and bring them together in a winning combination.

AFTERWORD

OUR SOCIETIES ARE BESET BY CHALLENGES – war, environmental crisis, population displacement, food shortages – yet these eight stories show human ingenuity and cooperation at their best. Thousands of scientists and physicians worked to solve some of the greatest health problems our species has faced, sometimes in partnership, sometimes in competition. Building on each other's efforts, they created tiny entities – molecules – that can impact patients with cancer, SMA, HIV, pain, sickle cell disease, Covid-19 and haemophilia. These stories represent the best of humanity.

Some people worry that drug discovery is running out of steam: we've harvested the 'low hanging fruit' of easily drugged targets, and we're reaching the limits of what can be done in medicine. I am convinced otherwise. I am enormously optimistic about the future. This is the Biological Century.

We've seen what humanity achieved in the Bronze Age, the Enlightenment, the Industrial Revolution, the Space Age and the Information Age. We are now in the Age of Biology. We've cracked many of the secrets of the genetic code and developed technologies to edit our own genes. We've learned how to image proteins and create artificial antibodies. We've mapped many of the biological complexities of the human body. We understand signalling pathways, growth factors, RNA splicing, blood coagulation, pathogen life cycles and foetal haemoglobin switching. We're learning more about our brains and immune systems and how the human genome relates to diseases. The best is yet to come, because the pace of innovation is accelerating in all these areas and more.

Some of the innovations in this book have taken decades to come to fruition. Nearly a hundred years passed between the insight that plasma contains blood clotting agents and the development of a bispecific antibody that mimics the function of Factor VIII.

Why has it all taken so long?

On the clinical side, we've had to describe and characterise diseases. We've had to develop tools to diagnose them, from stethoscopes to MRI scans, blood tests and more.

On the biology side, we've had to elucidate the mechanisms underlying the various disease states – decode nature. This required us to develop the fields of biochemistry, molecular biology, genetics and immunology among others.

On the technology side, we've had to build the pharmacologic tools to fix defects, stymie pathogens or 'restore nature'. That meant developing small molecules, antibodies, gene therapy, antisense oligonucleotides and so on.

And we've had to show, through years of exacting clinical trials, that each of these technologies is not only effective but also safe. To support this safety imperative, we've had to develop sophisticated toxicology studies and trial processes and networks, alongside a comprehensive regulatory framework.

In so doing, we've transformed a somewhat haphazard, even dangerous, drug discovery process into a global regulated research and development network that develops and tests more and more new molecules every year. The number of registered studies on ClinicalTrials.gov (a US government database of privately and publicly funded clinical studies conducted around the world) has exponentially increased from 1,255 in 2000 to 470,003 (including 186,387 involving a drug or biologic) as of 19 October 2023.

Humanity now has an extensive, international, experienced, well-funded, technologically advanced drug discovery machine, poised to take on our remaining health challenges – brain diseases, mental health conditions, autoimmune disorders, rare genetic diseases, emerging pathogens, antibiotic resistance and cancer among others. We can – and should – expect to deliver results much faster than we have in the past.

We've witnessed this acceleration of progress in our stories. Once the *PIK3CA* and *EGFR* mutations were found, alpelisib and osimertinib were developed much faster than earlier generations of cancer drugs. Nirmatrelvir was created and brought to the fight against Covid-19 in less than two years. When contrasted with the 1918–20 influenza pandemic, the overall Covid-19 pharmaceutical response was a vivid illustration of humanity's new capability and speed.

Imagine what can happen when we bring together novel chemistries, gene editing and antibody synthesis with cell engineering and next-generation biologics, all powered by big data, artificial intelligence, imaging technologies, and even more innovative clinical trial designs! That is why we can expect even more remarkable breakthroughs in the Biological Century.

As we see these breakthroughs come to fruition, let's remember that behind each one will be the stories of many unsung heroes who dedicated their lives to improving the health and well-being of humankind.

ACKNOWLEDGEMENTS

IF IT TAKES A VILLAGE TO make a drug, it has taken a small village to write this book.

First, I want to acknowledge the people interviewed for this book. Without them sharing their stories, there would be no book. I thank them for their time and thoughts: Loren Eng, Ayra Singh, Judith Melki, Adrian Krainer, Stuart Peltz, Luca Santerelli, Paulo Fontoura, Lutz Müller, Hasane Ratni, Mark Kris, Vince Miller, Thomas Lynch, Matthew Meyerson, Pasi Jänne, Serban Ghiorghiu, Darren Cross, Richard Ward, Andrew Mortlock, Teresa Klinowska, Susan Galbraith, Mark Anderton, Antoine Yver, Jean-Charles Soria, Kunihiro Hattori, Takehisa Kitazawa, Hiroyuki Saito, Tom Igawa, Gallia Levy, Midori Shima, Lew Cantley, Victor Velculescu, Sauveur-Michel Maira, Carlos Garcia-Echeverria, Christine Fritsch, Giorgio Caravatti, Pascal Furet, Cornelia Quadt, Celine Wilke, Michelle Miller, Jennifer Hammond, Annaliesa Anderson, Charlotte Allerton, Dafydd Owen, Matt Reese, Amit Kalgutkar, Angela Lukin, Melissa Avery, Sandeep Menon, Art Bergman, James Rusnak, Doug Higgs, Fyodor Urnov, Stuart Orkin, Vijay Sankaran, David Schlessinger, Jian Xu, Dan Bauer, David Altshuler, Samarth Kulkarni, Bill Hobbs, Bastiano Sanna, Wes Sundquist,

Tomas Cihlar, Christopher Aiken, Stephen Yant, John Link, Wade Blair, Roshy Pakdaman, Martin Rhee and Moupali Das.

Next, I want to thank the companies and executives who agreed to let us peek under the hood: Roche/Hans Clevers, AstraZeneca/Susan Galbraith, Chugai Pharmaceuticals/Hisafumi Yamada, Novartis/Jay Bradner and Fiona Marshall, Pfizer/Mikael Dolsten, Vertex/David Altshuler, CRISPR Therapeutics/Samarth Kulkarni, and Gilead/Merdad Persay. Importantly, thanks to their project/legal/communications leaders for shepherding the process: Lars Cleary, Kavita Patel, Heidi Bürgi and Daud Chaudry (Roche), Andrew White and Claire Nicholson (AstraZeneca), Tomoko Shimizu and Mari Otsuka (Chugai Pharmaceuticals), Sandra Schlüchter and Kara Cournoyer (Novartis), Niamh Slevin Roberts (Pfizer), Eleanor Celeste and Heather Nichols (Vertex), and Shirley Cantin and Brian Plummer (Gilead).

Thanks also to Harold Varmus and Katerina Politi for providing additional feedback on the osimertinib chapter. And thanks to Art Levinson and Adrian Hayday for allowing me to quote them.

I am grateful to Oneworld for their interest in publishing the book, and particularly my editor, Sam Carter.

Thanks to Euan Thorneycroft at A.M. Heath for representing the work to publishers.

Thanks to Hector Macdonald for conducting research including multiple interviews along with book consultation.

Thanks to Christine Micheel for fact-checking.

Finally, I would like to acknowledge the love and support of my family. To my sister, Maryland, and her husband, Steve, for encouraging me to pursue a life of medicine *and* science. And to my wife, Victoria, and my children, Allison and Lucas, for giving me the time to write this book. Special call out to Victoria, for her unwavering support of my desire to write a book about innovation in drug discovery, for stimulating me to think hard about innovation, and for insightful discussions along the way.

GLOSSARY

Acquired immunity Immunity that develops during a person's lifetime. There are two types of acquired immunity: active immunity and passive immunity. Active immunity develops after exposure to a disease-causing infectious microorganism or other foreign substance, such as following infection or vaccination. Passive immunity develops after a person receives immune system components, most commonly antibodies, from another person.[1]

Activator A class of substances that binds to, activates and increases the activity of a target.[2]

Adjuvant treatment Treatment using drugs, radiation or other means to stop cancer returning after surgery.

Animal model A non-human species used in biomedical research because it can mimic aspects of a biological process or disease found in humans.[3] Animal models are used to study the development and progression of diseases and to test new treatments before they are given to humans.[4]

Antibody In living organisms, a type of protein made by B lymphocytes in response to a foreign substance (antigen). Each naturally produced antibody only binds to a specific antigen, helping to destroy the antigen directly or assisting white blood cells to destroy the antigen.[5]

Bioavailability Rate and extent to which a drug is absorbed or is otherwise available to the treatment site in the body.[6]

Bispecific antibody An engineered antibody that binds to two target antigens, as compared to a conventional (monospecific) antibody which binds to one target antigen.

Blood stem cell Also known as haematopoietic stem cells. Blood cell development progresses from a haematopoietic stem cell, which can undergo either self-renewal or differentiation into a common lymphoid progenitor cell or a common myeloid progenitor cell. Red blood cells are derived from common myeloid cell progenitors.[7]

Breakthrough Therapy designation Breakthrough Therapy designation is a US Food and Drug Administration (FDA) process designed to expedite the development and review of drugs that are intended to treat a serious condition and preliminary clinical evidence indicates that the drug may demonstrate substantial improvement over available therapy on a clinically significant endpoint(s).[8]

Coagulation cascade The sequential process by which blood coagulation factors interact to form a fibrin clot.[9]

CRISPR/Cas9 A technique for gene editing that uses a guide DNA and Cas9 enzyme to cut DNA at a precise location. Also known as CRISPR gene editing.[10]

CYP enzyme CYP enzymes are found in the liver and are involved in drug metabolism.[11] Drug developers must pay close attention to how CYP enzymes interact with drug candidates, as metabolism that is too fast or too slow can be problematic.

Denaturation Occurs when biological molecules such as nucleic acids and proteins lose their functional structures. Can occur through use of physical (e.g. heat) or chemical (e.g. alcohol) means.

Discovery toxicology Toxicology is the study of the adverse effects of chemical, physical or biological agents on people, animals and the environment. Discovery toxicology is the focus of toxicology on preventing adverse effects of drugs by anticipating and addressing potentially problematic features of candidate molecules before initiation of clinical trials.[12]

Dose-response relationship A dose-response relationship occurs when administration of more drug leads to more effect, up to a maximum point.

Electroporation A procedure in which an electrical field is applied to cells for the purpose of increasing the permeability of the cell membrane, allowing DNA or other substances to be introduced into the cell.[13]

Enhancer A 50–150 base pair DNA sequence that increases the rate of transcription of coding sequences. It may be located at various distances and in either orientation upstream from, downstream from or within a structural gene.[14]

Epigenetics The study of changes affecting gene expression and its regulation without alteration of DNA sequence.[15]

Ex vivo Outside the living body; referring to the use or positioning of an organ, tissue or cell in an environment outside the living organism while the tissue or cells remain viable.[16]

Exon	The sequences of a gene that are present in the final, mature, spliced messenger RNA molecule from that gene.[17]
Formulation	The process of devising and assembling an active ingredient with other materials, such as solvents, fillers, binders, lubricants, preservatives and delivery vehicles (such as those needed for capsules, tablets, ointments etc.)
Genome-wide association study	A study that compares the complete DNA of people with a disease or condition to the DNA of people without the disease or condition. These studies find the genes involved in a disease, and may help prevent, diagnose and treat the disease. Also known as GWAS or whole genome association study.[18]
Guide RNA	A small RNA used in CRISPR/Cas9 gene editing. The guide RNA (gRNA) determines where the DNA cut occurs in the gene editing process.
Hepatic extraction	The process of the liver removing a drug from blood circulation.
Humanisation	The process of making an antibody drug candidate with protein sequences derived from animals more human-like and less likely to be attacked by the human immune system.
Hybridoma	Hybrid cells used in the *in vitro* production of specific monoclonal antibodies; produced by fusion of an established tissue culture line of plasma cells (e.g. mouse plasmacytoma cells) and specific antibody-producing cells (e.g. splenocytes from specifically immunised mice).[19]
In vitro	In the laboratory (outside the body). Referred to in contrast to *in vivo* (in the body).[20]
In vivo	Located or occurring in the body. Referred to in contrast to *in vitro* (in the laboratory).
International nonproprietary name (INN)	A unique name that is globally recognised and public property, which identifies pharmaceutical substances or active pharmaceutical ingredients. The INN name is established by the World Health Organization (WHO).[21]
Intron	The sequence of DNA in between exons that is initially copied into RNA but is spliced out of the final RNA transcript and therefore does not change the amino acid code. Some intronic sequences are known to affect gene expression.[22]
Investigational New Drug	A substance that has been tested in a laboratory and has got approval from the US Food and Drug Administration (FDA) to be tested in people. An experimental drug may be approved by the FDA for use in one disease or condition but be considered investigational in other diseases or conditions.[23]

Junk DNA Non-coding DNA corresponding to the portions of an organism's genome that do not code for amino acids, the building blocks of proteins. Some non-coding DNA sequences are known to serve functional roles, such as in the regulation of gene expression, while other areas of non-coding DNA currently have no known function. Non-coding DNA was previously referred to as junk DNA because scientists were unaware of any functions of non-coding DNA.[24]

Knock out Inactivation of specific genes by homologous recombination (a biological process involved in DNA repair). Knock outs are often created in laboratory organisms such as yeast or mice so that scientists can study what happens physiologically when a gene is removed (knocked out).[25]

Linkage analysis A gene-hunting technique that traces patterns of disease in high-risk families. It attempts to locate a disease-causing gene by identifying genetic markers of known chromosomal location that are co-inherited with a trait of interest.[26]

Lipinski's rule of five A concept frequently used in drug discovery to help predict if a biologically active molecule is likely to have the chemical and physical properties to be orally bioavailable.

Liquid biopsy Sampling and analysis of non-solid biological material (primarily blood), usually to determine if cancer cells or circulating-free cancer DNA is present.[27]

Mass balance study A study that looks at how an investigational drug is absorbed, distributed, metabolised and excreted in the human body.

Methaemoglobin A form of haemoglobin found in the blood in small amounts. Unlike normal haemoglobin, methaemoglobin cannot carry oxygen.[28]

Microtubule A component of the cell cytoskeleton that provides structure, shape and transport capabilities to cells. Microtubules are needed for cell division and thus are targets of anti-cancer drugs.

Mobilisation agents Drugs that cause blood stem cells to leave the bone marrow and circulate in the bloodstream. Such agents are used in the process of bone marrow transplantation.

Monoclonal antibody Antibodies are proteins naturally produced by B cells to help the immune system recognise antigens that cause disease, such as bacteria and viruses, and mark them for destruction. Monoclonal antibodies (mAbs) are antibody proteins made in laboratories to recognise specific targets. There are many mAb biologic drugs used in medicine.[29]

Monospecific antibody An antibody that targets one antigen.

Oncogene	A gene that normally directs cell growth, but that when mutated or overexpressed in a dominant fashion can release the cell from normal restraints on growth. The oncogene converts the cell into a tumour cell. Alterations can be inherited, or generated after birth, caused by, for example, environmental exposure to carcinogens.[30]
Oncogenesis	A pathologic process that involves the transformation of normal cells to a neoplastic state, resulting in cancerous cell proliferation. Also known as tumorigenesis.[31]
Oral bioavailability	Rate and extent to which an orally administered drug is absorbed through the gut and becomes available to the treatment site in the body.[32]
P24 capsid protein (CA protein)	The structural protein that forms the nucleocapsid of the human immunodeficiency virus.[33]
Pharmacokinetics	The characteristic movements of drugs within biological systems, as affected by absorption, distribution, binding, elimination, metabolism and excretion; particularly the rates of such movements.[34]
Phase I clinical trial	Phase I clinical trials are trials in which a new molecular entity (NME) is tested in humans for the first time. They are usually conducted in healthy volunteers, and the goal is to determine the NME's most frequent and serious adverse events, the NME's dose and schedule, and how the NME is broken down and excreted by the body. These trials usually involve a small number of participants. [35]
Phase II clinical trial	Phase II clinical trials gather preliminary data on whether a new molecular entity (NME) works in people who have a certain condition/disease (that is, the NME's effectiveness). For example, participants receiving the NME may be compared to similar participants receiving a different treatment, usually an inactive substance (called a placebo) or a standard of care drug. Safety and adverse events continue to be evaluated.[36]
Phase III clinical trial	Phase III clinical trials gather more information about a drug's safety and efficacy, ultimately with the aim of regulatory approval in a specific disease indication. These studies typically involve many participants. They are also known as pivotal or registrational trials.[37]
Phase IV studies	Phase IV studies include studies that are required by regulatory agencies post approval. These trials gather additional information about a drug's safety, efficacy or optimal use.[38]
Phosphorylation	A process in which a phosphate group (which consists of one phosphorus atom and four oxygen atoms) is added to a molecule, such as a protein.[39]
Pre-clinical	Refers to *in vitro* testing or *in vivo* animal testing before clinical trials in humans are to be carried out.[40]

Pre-clinical toxicology review The review of potential adverse events of a drug candidate before clinical trials in humans are to be carried out. Potential chemical, physical or biological causes of adverse events are evaluated.

Prodrug A prodrug is a compound that metabolises into the active drug in the body following administration.

Protease inhibitor Any substance that inhibits a protease, an enzyme that breaks peptide bonds.[41]

Quadroma Hybrid cells used historically in the *in vitro* production of bispecific antibodies. Also called a hybrid hybridoma to highlight its possible origin as a fusion of two hybridoma cell lines that produce specific monoclonal antibodies.[42]

Recombinant Made by combining genetic material from two different sources.[43]

Replacement factor Treatment of haemophilia by replacing the missing coagulation factor: Factor VIII for people with haemophilia A; Factor IX for those with haemophilia B.

Repressor A protein that results in decreased expression of some gene product.

Retrovirus The class of viruses that can integrate some of their genetic code into their host cell's DNA.

RNA splicing The process by which introns, the noncoding regions of genes, are excised out of the primary messenger RNA transcript, and the exons (i.e. coding regions) are joined together to generate mature messenger RNA. The latter serves as the template for synthesis of a specific protein.[44]

Silencing Occurs when normal production of a protein is stopped, through interruption of transcription or translation.

Single-nucleotide polymorphisms (SNPs) DNA sequence variations that occur when a single nucleotide (adenine, thymine, cytosine or guanine) in the genome sequence is altered; usually present in at least one per cent of the population; pronounced snip.[45]

Surface polarity The degree to which the exterior surface of a molecule is free to interact via dipole-dipole interactions or hydrogen bonds with other nearby molecules. Polarity is caused by differences in electronegativity (how strongly the nuclei attract the electrons) between atoms bonded to one another, and surface polarity is determined by the polarity of the chemical bonds closest to the surface of a molecule. This concept is most often considered relative to large molecules such as proteins that have an exterior surface and an interior inaccessible to other molecules.

Toxicology Toxicology is the study of the adverse effects of chemical, physical or biological agents on people, animals and the environment.[46]

Transcription The process of synthesising messenger RNA (mRNA) from DNA.[47]

Transcription factor Transcription factors are a diverse group of proteins that bind to DNA at specific promoter or enhancer regions. They also bind to DNA-associated proteins to initiate, stimulate, inhibit or terminate transcription.[48]

Transgenic Characterised by transfer of a gene or genes from another species or breed.[49]

Transposon Sequences of DNA that can seemingly 'jump' from one part of the genome to another.

Viral envelope The outer structure that encloses the nucleocapsids of some viruses.[50]

Virion A complete viral particle capable of persisting outside of a host cell. In the case of HIV, consists of the viral envelope, capsid and genetic material.

Wildtype The naturally occurring, normal, non-mutated version of a gene or genome.[51]

X-ray crystallography A technique wherein an image of the arrangement of atoms in a crystal is constructed from the pattern produced by the diffraction of X-rays through the closely spaced lattice of atoms in a crystal.[52]

References

Definitions have been sourced as much as possible from US government resources. Most US government texts are free of copyright and may be used without permission; citation is permitted. In this glossary, I have used or edited definitions from several of these resources, including the National Cancer Institute Thesaurus (which itself provides definitions from multiple other resources within the US Department of Health and Human Services), the Glossary of HIV/AIDS-related terms, the National Human Genome Research Institute's Talking Glossary of Genomic and Genetic Terms, the Drugs@FDA Glossary of Terms, and the ClinicalTrials.gov Glossary Terms, and others.

LIST OF FIGURES AND TABLES

Figures

Tables

NOTES

Introduction

1. Rocha e Silva M., Beraldo W.T., Rosenfeld G. 'Bradykinin, a hypotensive and smooth muscle stimulating factor released from plasma globulin by snake venoms and by trypsin.' *Am J Physiol.* 1949 Feb; 156(2): 261–73.
2. Downey, P. 'Profile of Sérgio Ferreira.' *PNAS.* 2008 Dec; 105(49): 19035–7.
3. Ferreira S.H. 'A bradykinin-potentiating factor (bpf) present in the venom of *Bothrops jararca.*' *British Journal of Pharmacology and Chemotherapy.* 1965 Feb; 24(1): 163–9.
4. https://www.nobelprize.org/prizes/medicine/1982/vane/biographical/
5. https://qmro.qmul.ac.uk/xmlui/bitstream/handle/123456789/22586/e2017125.pdf
6. Riordan J.F. 'Angiotensin-I-converting enzyme and its relatives.' *Genome Biology.* 2003; 4(8): 225.
7. https://www.who.int/news-room/fact-sheets/detail/cancer#

1: The World's Most Common Rare Disease

1. Darras B.T. 'Spinal muscular atrophies.' *Pediatric Clinics of North America.* 2015 Jun; 62(3): 743–66.
2. Cartegni, L., Krainer, A. 'Correction of disease-associated exon skipping by synthetic exon-specific activators.' *Nature Structural & Molecular Biology* 10. 2003. 120–5.

3. Hua Y., Vickers T.A., Okunola H.L. et al. 'Antisense masking of an hnRNP A1/A2 intronic splicing silencer corrects SMN2 splicing in transgenic mice.' *American Journal of Human Genetics*. 2008 Apr; 82(4): 834–48.
4. Naryshkin N.A., Weetall M., Dakka A. et al. 'Motor neuron disease. SMN2 splicing modifiers improve motor function and longevity in mice with spinal muscular atrophy.' *Science*. 2014 Aug 8; 345(6197): 688–93.
5. Kletzl H., Marquet A., Günther A. et al. 'The oral splicing modifier RG7800 increases full length survival of motor neuron 2 mRNA and survival of motor neuron protein: Results from trials in healthy adults and patients with spinal muscular atrophy.' *Neuromuscular Disorders*. 2019 Jan; 29(1): 21–9.
6. https://www.nationwidechildrens.org/newsroom/news-releases/2013/10/avexis-biolife-licenses-spinal-muscular-atrophy-sma-patent-portfolio-from-nationwide-childrens; Mendell J.R., Al-Zaidy S., Shell R. et al. 'Single-Dose Gene-Replacement Therapy for Spinal Muscular Atrophy.' *New England Journal of Medicine*. 2017 Nov 2; 377(18): 1713–22.

2: Lung Cancer in Never Smokers

1. https://www.nobelprize.org/womenwhochangedscience/stories/rita-levi-montalcini
2. https://www.nytimes.com/1988/05/01/books/a-self-made-scientist.html
3. https://www.ncbi.nlm.nih.gov/pmc/articles/PMC3612650/pdf/pnas.201301976.pdf
4. https://www.nytimes.com/1988/05/01/books/a-self-made-scientist.html
5. https://doi.org/10.1016/S0140-6736(20)30550-X
6. https://www.science.org/doi/10.1126/science.abb4095
7. Zeng F., Harris R.C. 'Epidermal growth factor, from gene organization to bedside.' *Seminars in Cell and Developmental Biology*. 2014 Apr; 28: 2–11.
8. https://www.nobelprize.org/uploads/2018/06/cohen-lecture.pdf; https://news.vanderbilt.edu/2011/12/09/stanley-cohen-nobel-prize/
9. https://doi.org/10.1016/S0140-6736(20)30550-X
10. https://news.vanderbilt.edu/2011/12/09/stanley-cohen-nobel-prize
11. Merlino G.T., Xu Y.H., Ishii S. et al. 'Amplification and enhanced expression of the epidermal growth factor receptor gene in A431 human carcinoma cells.' *Science*. 1984 Apr 27; 224(4647): 417–9.
12. Downward J., Yarden Y., Mayes E. et al. 'Close similarity of epidermal growth factor receptor and v-erb-B oncogene protein sequences.' *Nature*. 1984 Feb 9–15; 307(5951): 521–7.
13. Wong A.J., Ruppert J.M., Bigner S.H. et al. 'Structural alterations of the epidermal growth factor receptor gene in human gliomas.' *Proceedings of the National Academy of Sciences (PNAS)*. 1992 Apr 1; 89(7): 2965–9.
14. https://doi.org/10.1016/S0140-6736(20)30550-X

15. Pao W., Miller V., Zakowski M. et al. 'EGF receptor gene mutations are common in lung cancers from "never smokers" and are associated with sensitivity of tumors to gefitinib and erlotinib.' *Proceedings of the National Academy of Sciences (PNAS)*. 2004 Sep 7; 101(36): 13306–11.
16. Kalanithi, Paul. *When Breath Becomes Air* (New York: Random House, 2016).
17. https://med.stanford.edu/news/all-news/2015/03/stanford-neurosurgeon-writer-paul-kalanithi-dies-at-37.html
18. Kobayashi S., Boggon T.J., Dayaram T. et al. 'EGFR mutation and resistance of non-small-cell lung cancer to gefitinib.' *New England Journal of Medicine*. 2005 Feb 24; 352(8): 786–92.
19. Pao W., Miller V.A., Politi K.A. et al. 'Acquired resistance of lung adenocarcinomas to gefitinib or erlotinib is associated with a second mutation in the EGFR kinase domain.' *PLOS Medicine*. 2005 Mar; 2(3): e73.
20. Mok T.S., Wu Y.L., Thongprasert S. et al. 'Gefitinib or carboplatin-paclitaxel in pulmonary adenocarcinoma.' *New England Journal of Medicine*. 2009 Sep 3; 361(10): 947–57.
21. http://archive.boston.com/news/nation/articles/2003/11/24/a_drug_that_works____for_some/
22. Lynch T.J., Bell D.W., Sordella R. et al. 'Activating mutations in the epidermal growth factor receptor underlying responsiveness of non-small-cell lung cancer to gefitinib.' *New England Journal of Medicine*. 2004 May 20; 350(21): 2129–39.
23. Paez J.G., Jänne P.A., Lee J.C. et al. 'EGFR mutations in lung cancer: correlation with clinical response to gefitinib therapy.' *Science*. 2004 Jun 4; 304(5676): 1497–500.
24. Shepherd F.A., Rodrigues Pereira J., Ciuleanu T. et al. 'National Cancer Institute of Canada Clinical Trials Group. Erlotinib in previously treated non-small-cell lung cancer.' *New England Journal of Medicine*. 2005 Jul 14; 353(2): 123–32.
25. Cross D., Alessi D., Cohen P. et al. 'Inhibition of glycogen synthase kinase-3 by insulin mediated by protein kinase B.' *Nature*. 1995; 378: 785–9.
26. Kettle J.G., Wilson D.M. 'Standing on the shoulders of giants: a retrospective analysis of kinase drug discovery at AstraZeneca.' *Drug Discovery Today*. 2016 Oct; 21(10): 1596–608.
27. Sacher A.G., Jänne P.A., Oxnard G.R. 'Management of acquired resistance to epidermal growth factor receptor kinase inhibitors in patients with advanced non-small cell lung cancer.' *Cancer*. 2014 Aug 1; 120(15): 2289–98.
28. Cross D.A., Ashton S.E., Ghiorghiu S. et al. 'AZD9291, an irreversible EGFR TKI, overcomes T790M-mediated resistance to EGFR inhibitors in lung cancer.' *Cancer Discovery*. 2014 Sep; 4(9): 1046–61.
29. Jänne P.A., Yang J.C., Kim D.W. et al. 'AZD9291 in EGFR inhibitor-resistant non-small-cell lung cancer.' *New England Journal of Medicine*. 2015 Apr 30; 372(18): 1689–99.

30. Cross D.A., Ashton S.E., Ghiorghiu S. et al. 'AZD9291, an irreversible EGFR TKI, overcomes T790M-mediated resistance to EGFR inhibitors in lung cancer.' *Cancer Discovery*. 2014 Sep; 4(9): 1046–61.
31. Soria J.C., Ohe Y., Vansteenkiste J. et al. 'Osimertinib in Untreated EGFR-Mutated Advanced Non-Small-Cell Lung Cancer.' *New England Journal of Medicine*. 2018 Jan 11; 378(2): 113–125.
32. Herbst R., Tsuboi M., John T. et al. 'Overall survival analysis from the ADAURA trial of adjuvant osimertinib in patients with resected EGFR-mutated (EGFRm) stage IB–IIIA non-small cell lung cancer (NSCLC).' *Journal of Clinical Oncology*. 2023 41: 17_suppl, LBA3-LBA3.

3: The Universal Affliction

1. 'Ein glücklicher Zufall hat uns ein Präparat in die Hand gespielt.' https://link.springer.com/referenceworkentry/10.1007/978-3-642-17907-5_2; https://archive.org/details/s4366id1397390/page/560/mode/2up
2. https://journals.sagepub.com/doi/pdf/10.1177/0310057X1103900301
3. https://www.kallegroup.com/en/company/history/
4. https://americanhistory.si.edu/collections/search/object/nmah_720128#
5. 'Toxic effects of antifebrin.' *JAMA*. 1888; X(22): 688.
6. Brune K., Renner B., Tiegs G. 'Acetaminophen/paracetamol: A history of errors, failures and false decisions.' *European Journal of Pain*. 2015 Aug; 19(7): 953–65.
7. Armstrong, H. 'Chemical Industry and Carl Duisberg.' *Nature*. 1935; 135: 1021–5.
8. Konantz, W. A. 'The manufacture of acetphenetidin.' *Journal of the American Pharmaceutical Association*. 1919; 8: 284–90.
9. https://cite.case.law/f/108/233/
10. Brune K., Renner B., Tiegs G. 'Acetaminophen/paracetamol: A history of errors, failures and false decisions.' *European Journal of Pain*. 2015 Aug; 19(7): 953–65.
11. Margetts G. 'Phenacetin and Paracetamol.' *Journal of International Medical Research*. 1976; 4(4_suppl): 55–70.
12. https://cite.case.law/f/108/233/
13. Tan G.H., Rabbino M.D., Hopper J. Jr. 'Is phenacetin a nephrotoxin? A report on twenty-three users of the drug.' *California Medicine*. 1964 Aug; 101(2): 73–7; Ramsay A.G., White D.F. 'Phenacetin nephropathy.' *Canadian Medical Association Journal*. 1965 Jan 9; 92(2): 55–9.
14. Margetts G. 'Phenacetin and Paracetamol.' *Journal of International Medical Research*. 1976; 4(4_suppl): 55–70.
15. Iversen L. 'Julius Axelrod. 30 May 1912–29 December 2004.' *Biographical Memoirs of Fellows of the Royal Society*. 2006; 52: 1–13.

16. Halperin E.C. 'Why Did the United States Medical School Admissions Quota for Jews End?' *American Journal of Medical Science*. 2019 Nov; 358(5): 317–25.
17. Axelrod, J. 'An Unexpected Life in Research.' *Annual Review of Pharmacology and Toxicology*. 1988; 28(1): 1–24.
18. Ibid.
19. Brodie B.B., Axelrod J. 'The Fate of Acetanilide in Man.' *Journal of Pharmacology and Experimental Therapeutics*. 1948; 94: 29–38.
20. Lester D., Greenberg L.A. 'The metabolic fate of acetanilid and other aniline derivatives; major metabolites of acetanilid appearing in the blood.' *Journal of Pharmacology and Experimental Therapeutics*. 1947 May; 90(1): 68–75.
21. Brodie B.B., Axelrod J. 'The fate of acetophenetidin in man and methods for the estimation of acetophenetidin and its metabolites in biological material.' *Journal of Pharmacology and Experimental Therapeutics*. 1949 Sep; 97(1): 58–67.
22. Bennett C.L., Starko K.M., Thomsen H.S. et al. 'Linking drugs to obscure illnesses: lessons from pure red cell aplasia, nephrogenic systemic fibrosis, and Reye's syndrome. A report from the Southern Network on Adverse Reactions (SONAR).' *Journal of General Internal Medicine*. 2012 Dec; 27(12): 1697–703.
23. Kulich R., Loeser J.D. 'The Business of Pain Medicine: The Present Mirrors Antiquity.' *Pain Medicine*. 2011 July; 12(7): 1063–75.
24. https://www.latimes.com/archives/la-xpm-2010-may-31-la-me-robert-mcneil-20100531-story.html
25. https://www.kilmerhouse.com/2009/12/1959-mcneil-laboratories-joins-the-family
26. Kulich R., Loeser J.D. 'The Business of Pain Medicine: The Present Mirrors Antiquity.' *Pain Medicine*. 2011 July; 12(7): 1063–75.
27. https://www.latimes.com/archives/la-xpm-2010-may-31-la-me-robert-mcneil-20100531-story.html
28. https://edu.rsc.org/feature/pain-relief-from-coal-tar-to-paracetamol/2020140.article
29. https://www.latimes.com/archives/la-xpm-2010-may-31-la-me-robert-mcneil-20100531-story.html
30. https://profiles.nlm.nih.gov/spotlight/hh/feature/biographical-overview
31. Axelrod, J. 'An Unexpected Life in Research.' *Annual Review of Pharmacology and Toxicology*. 1988; 28(1): 1–24.

4: The Royal Disease

1. https://history.rcplondon.ac.uk/inspiring-physicians/robert-gwyn-macfarlane
2. Hougie C. 'The waterfall-cascade and authoprothombin hypotheses of blood coagulation: personal reflections from an observer.' *Journal of Thrombosis and Haemostasis*. 2004; 2: 1225–33.

3. Macfarlane R.G. 'Russell's viper venom,' 1934–64. *British Journal of Haematology*. 1967 Jul; 13(4): 437–51.
4. Macfarlane R.G. 'An enzyme cascade in the blood clotting mechanism, and its function as a biochemical amplifier.' *Nature*. 1964 May 2; 202: 498–9.
5. White G.C. 'Hemophilia: an amazing 35-year journey from the depths of HIV to the threshold of cure.' *Transactions of the American and Climatological Association*. 2010; 121: 61–73; discussion 74–5.
6. Kasper C.K. 'Judith Graham Pool and the discovery of cryoprecipitate.' *Haemophilia*. 2012 Nov; 18(6): 833–5.
7. White G.C. 'Hemophilia: an amazing 35-year journey from the depths of HIV to the threshold of cure.' *Transactions of the American and Climatological Association*. 2010; 121: 61–73; discussion 74–5.
8. Vehar G., Keyt B., Eaton D. et al. 'Structure of human factor VIII.' *Nature*. 1984; 312, 337–42; Ling G., Tuddenham E.G.D. 'Factor VIII: the protein, cloning its gene, synthetic factor and now – 35 years later – gene therapy; what happened in between?' *British Journal of Haematology*. 2020 May; 189(3): 400–7.
9. https://pink.pharmaintelligence.informa.com/PS021887/BAXTERGENETICS-INSTITUTEs-RECOMBINATE-ANTIHEMOPHILIC-FACTOR-APPROVED-BY-FDA-DEC-10-FIRST-LICENSED-rFACTOR-VIII-PRODUCT-WILL-BEGIN-SHIPPING-DEC-21
10. https://pink.pharmaintelligence.informa.com/PS022223/MILES-KOGENATE-SHIPMENTS-MAY-BEGIN-AS-EARLY-AS-WEEK-OF-MARCH-1
11. Riethmüller G. 'Symmetry breaking: bispecific antibodies, the beginnings, and 50 years on.' *Cancer Immunity*. 2012; 12: 12.
12. Nisonoff A. et al. 'Properties of the Major Component of a Peptic Digest of Rabbit Antibody.' *Science*. 1960; 132: 1770–1.
13. https://www.nobelprize.org/prizes/medicine/1984/press-release/
14. Milstein C., Cuello A.C. 'Hybrid hybridomas and their use in immunohistochemistry.' *Nature*. 1983 Oct 6–12; 305(5934): 537–40.
15. Staerz U.D., Kanagawa O., Bevan M.J. 'Hybrid antibodies can target sites for attack by T cells.' *Nature*. 1985 Apr 18–24; 314(6012): 628–31.
16. Riethmüller G. 'Symmetry breaking: bispecific antibodies, the beginnings, and 50 years on.' *Cancer Immunity*. 2012; 12: 12.
17. Merchant A.M., Zhu Z., Yuan J.Q. et al. 'An efficient route to human bispecific IgG.' *Nature Biotechnology*. 1998 Jul; 16(7): 677–81.
18. Kitazawa T., Igawa T., Sampei Z. et al. 'A bispecific antibody to factors IXa and X restores factor VIII hemostatic activity in a hemophilia A model.' *Nature Medicine*. 2012 Oct; 18(10): 1570–4.
19. Ibid.; Sampei Z., Igawa T., Soeda T. et al. 'Identification and multidimensional optimization of an asymmetric bispecific IgG antibody mimicking the function of factor VIII cofactor activity.' *PLOS ONE*. 2013; 8(2): e57479.
20. Uchida N., Sambe T., Yoneyama K. et al. 'A first-in-human phase 1 study of ACE910, a novel factor VIII-mimetic bispecific antibody, in healthy subjects.' *Blood*. 2016 Mar 31; 127(13): 1633–41.

21. Shima M., Hanabusa H., Taki M. et al. 'Long-term safety and efficacy of emicizumab in a phase 1/2 study in patients with hemophilia A with or without inhibitors.' *Blood Advances.* 2017 Sep 27; 1(22): 1891–9.
22. https://www.fda.gov/drugs/resources-information-approved-drugs/fda-approves-emicizumab-kxwh-prevention-and-reduction-bleeding-patients-hemophilia-factor-viii

5: Therapy-Resistant Breast Cancer

1. Vasella D., Slater R. *Magic Cancer Bullet: How a Tiny Orange Pill May Rewrite Medical History* (New York: HarperCollins, 2004).
2. Velculescu V.E., Zhang L., Vogelstein B. et al. 'Serial analysis of gene expression.' *Science.* 1995 Oct 20; 270(5235): 484–7.
3. Samuels Y., Wang Z., Bardelli A. et al. 'High frequency of mutations of the PIK3CA gene in human cancers.' *Science.* 2004 Apr 23; 304(5670): 554.
4. Ibid.
5. Furet P., Guagnano V., Fairhurst R.A. et al. 'Discovery of NVP-BYL719 a potent and selective phosphatidylinositol-3 kinase alpha inhibitor selected for clinical evaluation.' *Bioorganic & Medicinal Chemistry Letters.* 2013 Jul 1; 23(13): 3741–8.
6. https://www.novartis.com/news/media-releases/novartis-piqray-data-show-survival-benefit-patients-hrher2-advanced-breast-cancer-pik3ca-mutation

6: The Virus That Stopped the World

1. Millar J., Şahin U., Türeci Ö. *The Vaccine: Inside the Race to Conquer the Covid-19 Pandemic* (London: Welbeck, 2021); Gilbert S., Green C. *Vaxxers: A Pioneering Moment in Scientific History* (London: Hodder & Stoughton, 2021); Bourla A. *Moonshot: Inside Pfizer's Nine-Month Race to Make the Impossible Possible* (New York: HarperBusiness, 2022).
2. https://www.cdc.gov/sars/about/fs-sars.html
3. Collins C. et al.,'The NIH-led research response to COVID-19.' *Science.* 2023; 379: 441–4.
4. https://www.wsj.com/articles/new-york-citys-deadliest-day-from-covid-19-hit-one-year-ago-11617796817
5. https://www.pfizer.com/sites/default/files/investors/financial_reports/annual_reports/2021/story/new-course-for-covid-treatment/
6. https://covid19.who.int/
7. https://www.pfizer.com/news/press-release/press-release-detail/pfizers-paxlovidtm-receives-fda-approval-adult-patients

8. Haller D.A., Colwell R.C. 'The protective qualities of the gauze face mask: experimental studies.' *JAMA*. 1918; 71(15): 1213–5.
9. https://covid19.who.int/
10. Msemburi W., Karlinsky A., Knutson V. et al. 'The WHO estimates of excess mortality associated with the COVID-19 pandemic.' *Nature*. 2023; 613: 130–7.
11. Barry J.M. *The Great Influenza: The Story of the Deadliest Pandemic in History* (New York: Penguin, 2009).

7: The Dawn of Gene Editing

1. Steensma D.P., Kyle R.A., Shampo M.A. 'Walter Clement Noel – first patient described with sickle cell disease.' *Mayo Clinic Proceedings*. 2010 Oct; 85(10): e74–5.
2. http://huhealthcare.com/healthcare/hospital/specialty-services/sickle-cell-disease-center/disease-information/
3. https://www.nytimes.com/2020/01/11/health/sickle-cell-disease-cure.html
4. https://www.thelancet.com/journals/lancet/article/PIIS0140-6736(14)60635-8/fulltext
5. Spinney L. 'Anthony Allison.' *BMJ*. 2014; 348: g2243.
6. Allison A.C. 'The Discovery of Resistance to Malaria of Sickle-cell Heterozygotes.' *Biochemistry and Molecular Biology Education*. 2002; 30(5): 279–87.
7. Haldane J.B.S. 'The rate of mutation of human genes.' *Hereditas*. 1949; 35: 267–73.
8. Allison A.C. 'The Discovery of Resistance to Malaria of Sickle-cell Heterozygotes.' *Biochemistry and Molecular Biology Education*. 2002; 30(5): 279–87.
9. Pauling L., Itano H.A. et al. 'Sickle cell anemia a molecular disease.' *Science*. 1949 Nov 25; 110(2865): 543–8.
10. Ingram V.M. 'A specific chemical difference between the globins of normal human and sickle-cell anaemia haemoglobin.' *Nature*. 1956 Oct 13; 178(4537): 792–4.
11. Watson J. 'The significance of the paucity of sickle cells in newborn Negro infants.' *American Journal of Medical Science*. 1948; 215(4): 419–23.
12. https://newsroom.uw.edu/postscript/george-stamatoyannopoulos-blood-research-pioneer-dies-84
13. https://ashpublications.org/thehematologist/article/doi/10.1182/hem.V15.4.8795/463011/George-Stamatoyannopoulos-MD-1934-2018-The-Passing
14. Srivastava A., Kay M.A., Athanasopoulos T. et al. 'A Tribute to George Stamatoyannopoulos.' *Human Gene Therapy*. 2016 Apr; 27(4): 280–6.
15. Fessas P., Stamatoyannopoulos G. 'Hereditary persistence of fetal hemoglobin in Greece. A study and a comparison.' *Blood*. 1964; 24(3): 223–40.

16. Ibid.; https://newsroom.uw.edu/postscript/george-stamatoyannopoulos-blood-research-pioneer-dies-84
17. https://www.nytimes.com/2020/01/11/health/sickle-cell-disease-cure.html
18. Kazazian H.H. Jr. '2014 William Allan Award introduction: Stuart Orkin.' *American Journal of Human Genetics*. 2015 Mar 5; 96(3): 352–3.
19. https://medlineplus.gov/genetics/understanding/genomicresearch/snp/
20. Cousens N.E., Gaff C.L., Metcalfe S.A. et al. 'Carrier screening for beta-thalassaemia: a review of international practice.' *European Journal of Human Genetics*. 2010 Oct; 18(10): 1077–83.
21. 'Obituary.' *L'Unione Sarda*. 19 April 2005: 38.
22. Pilia G., Chen W.M., Scuteri A. et al. 'Heritability of cardiovascular and personality traits in 6,148 Sardinians.' *PLOS Genetics*. 2006 Aug 25; 2(8): e132.
23. https://www.nytimes.com/2020/01/11/health/sickle-cell-disease-cure.html
24. Menzel S., Garner C., Gut I. et al. 'A QTL influencing F cell production maps to a gene encoding a zinc-finger protein on chromosome 2p15.' *Nature Genetics*. 2007 Oct; 39(10): 1197–9.
25. Uda M., Galanello R., Sanna S. et al. 'Genome-wide association study shows BCL11A associated with persistent fetal hemoglobin and amelioration of the phenotype of beta-thalassemia.' *Proceedings of the National Academy of Sciences of the United States of America*. 2008 Feb 5; 105(5): 1620–5.
26. Sankaran V.G., Menne T.F., Xu J. et al. 'Human fetal hemoglobin expression is regulated by the developmental stage-specific repressor BCL11A.' *Science*. 2008 Dec 19; 322(5909): 1839–42.
27. Sankaran V.G., Xu J., Ragoczy T. et al. 'Developmental and species-divergent globin switching are driven by BCL11A.' *Nature*. 2009 Aug 27; 460(7259): 1093–7.
28. Xu J., Peng C., Sankaran V.G. et al. 'Correction of sickle cell disease in adult mice by interference with fetal hemoglobin silencing.' *Science*. 2011 Nov 18; 334(6058): 993–6.
29. https://www.jccfund.org/blog/elenoe-crew-smith/
30. https://www.asbmb.org/asbmb-today/industry/040116/devoted-to-dna
31. Vierstra J., Reik A., Chang K.H. et al. 'Functional footprinting of regulatory DNA.' *Nature Methods*. 2015; 12: 927–30.
32. Bauer D.E., Canver M.C., Smith E.C., et al. 'Crispr-Cas9 Saturating Mutagenesis Reveals an Achilles Heel in the BCL11A Erythroid Enhancer for Fetal Hemoglobin Induction (by Genome Editing).' *Blood*. 2015; 126 (23): 638.
33. For example, Isaacson W. *The Code Breaker: Jennifer Doudna, Gene Editing, and the Future of the Human Race* (New York: Simon & Schuster, 2021).
34. https://www.npr.org/sections/health-shots/2020/12/15/944184405/1st-patients-to-get-crispr-gene-editing-treatment-continue-to-thrive
35. The FDA gave approval for CTX001 to be used to treat sickle cell patients on 8 December 2023 https://www.fda.gov/news-events/press-announcements/fda-approves-first-gene-therapies-treat-patients-sickle-cell-disease.
36. https://news.vrtx.com/news-releases/news-release-details/positive-results

-pivotal-trials-exa-cel-transfusion-dependent; Frangoul H., Locatelli F., Bhatia M. et al. 'Efficacy and Safety of a Single Dose of Exagamglogene Autotemcel for Severe Sickle Cell Disease.' *Blood*. 2022; 140 (Supplement 1): 29–31.

8: Viral Geometry

1. De Cock K.M., Jaffe H.W., Curran J.W. 'Reflections on 30 years of AIDS.' *Emerging Infectious Diseases*. 2011 Jun; 17(6): 1044–8.
2. https://www.youtube.com/watch?v=7Kp1PKENhpA
3. https://www.nobelprize.org/prizes/medicine/2008/summary/
4. https://www.washingtonpost.com/archive/politics/1998/02/04/aids-virus-identified-in-blood-sample-taken-from-african-man-in-1959/ccb2375b-a150-473d-8f5d-f60669b911ad/
5. https://time.com/4705809/first-aids-drug-azt/
6. Gitti R.K., Lee B.M., Walker J. et al. 'Structure of the amino-terminal core domain of the HIV-1 capsid protein.' *Science*. 1996; 273(5272): 231–5; Gamble T.R., Vajdos F.F., Yoo S. et al. 'Crystal structure of human cyclophilin A bound to the amino-terminal domain of HIV-1 capsid.' *Cell*. 1996; 87(7): 1285–94; Gamble T.R., Yoo S., Vajdos F.F. et al. 'Structure of the carboxyl-terminal dimerization domain of the HIV-1 capsid protein.' *Science*. 1997; 278(5339): 849–53.
7. Li S., Hill C., Sundquist W. et al. 'Image reconstructions of helical assemblies of the HIV-1 CA protein.' *Nature*. 2000; 407: 409–13.
8. von Schwedler U.K., Stray K.M., Garrus J.E. et al. 'Functional surfaces of the human immunodeficiency virus type 1 capsid protein.' *Journal of Virology*. 2003 May; 77(9): 5439–50.
9. https://ourworldindata.org/hiv-aids
10. https://data.unaids.org/topics/young-people/childrenonthebrink_en.pdf
11. https://www.niaid.nih.gov/diseases-conditions/antiretroviral-drug-development
12. d'Arminio Monforte A., Lepri A.C., Rezza G. et al. 'Insights into the reasons for discontinuation of the first highly active antiretroviral therapy (HAART) regimen in a cohort of antiretroviral naïve patients. I.C.O.N.A. Study Group. Italian Cohort of Antiretroviral-Naïve Patients.' *AIDS*. 2000; 14: 499–507; Lucas G.M., Chaisson R.E., Moore R.D. 'Highly active antiretroviral therapy in a large urban clinic: risk factors for virologic failure and adverse drug reactions.' *Annals of Internal Medicine*. 1999; 131: 81–7.
13. Forshey B.M., von Schwedler U., Sundquist W.I. et al. 'Formation of a human immunodeficiency virus type 1 core of optimal stability is crucial for viral replication.' *Journal of Virology*. 2002 Jun; 76(11): 5667–77.
14. Tang C., Loeliger E., Kinde I. et al. 'Antiviral inhibition of the HIV-1 capsid protein.' *Journal of Molecular Biology*. 2003 Apr 11; 327(5): 1013–20.

15. Ehrlich L.S., Agresta B.E., Carter C.A. 'Assembly of recombinant human immunodeficiency virus type 1 capsid protein in vitro.' *Journal of Virology*. 1992; 66: 4874–83.
16. Hung M., Niedziela-Majka A., Jin D. et al. 'Large-Scale Functional Purification of Recombinant HIV-1 Capsid.' *PLOS ONE*. 2013 Mar; 158(3): e58035.
17. Scott D., Bayly A., Abell C. et al. 'Small molecules, big targets: drug discovery faces the protein–protein interaction challenge.' *Nature Reviews Drug Discovery*. 2016; 15: 533–50.
18. Pornillos O., Ganser-Pornillos B.K., Kelly B.N. et al. 'X-ray structures of the hexameric building block of the HIV capsid.' *Cell*. 2009 Jun 26; 137(7): 1282–92.
19. Blair W.S., Pickford C., Irving S.L. et al. 'HIV capsid is a tractable target for small molecule therapeutic intervention.' *PLOS Pathogens*. 2010 Dec 9; 6(12): e1001220.
20. Appleby T.C., Link J.O., Yant S.R. et al. 'Crystal structure of the HIV capsid hexamer bound to the small molecule long-acting inhibitor, GS-6207.' *RCSB Protein Data Bank*. 2019 Nov 11: https://www.rcsb.org/structure/6v2f
21. Link J.O., Rhee M.S., Tse W.C. et al. 'Clinical targeting of HIV capsid protein with a long-acting small molecule.' *Nature*. 2020 Aug; 584(7822): 614–8.
22. Yant S.R., Mulato A., Hansen D. et al. 'A highly potent long-acting small-molecule HIV-1 capsid inhibitor with efficacy in a humanized mouse model.' *Nature Medicine*. 2019 Sep; 25(9): 1377–84.
23. https://www.sunlencahcp.com/how-sunlenca-works/first-in-class-moa/
24. https://cen.acs.org/pharmaceuticals/drug-discovery/Wrestling-Lipinski-rule-5/101/i8
25. Begley R., Lutz J., Rhee M. et al. 'GS-6207 sustained delivery formulation supports 6-month dosing interval.' *2020 International AIDS Conference*; Virtual; July 6–10, 2020. Abstract PEB0265: https://programme.aids2020.org/Abstract/Abstract/8533
26. Dvory-Sobol H., Shaik N., Callebaut C. et al. 'Lenacapavir: a first-in-class HIV-1 capsid inhibitor.' *Current Opinion in HIV and AIDS*. 2022 Jan; 17(1): 15–21.
27. https://www.gilead.com/news-and-press/press-room/press-releases/2020/11/gilead-announces-investigational-longacting-hiv1-capsid-inhibitor-lenacapavir-achieves-primary-endpoint-in-phase-23-study-in-heavily-treatmentex; Segal-Maurer S., DeJesus E., Stellbrink H.J. et al. 'Capsid Inhibition with Lenacapavir in Multidrug-Resistant HIV-1 Infection.' *New England Journal of Medicine*. 2022 May 12; 386(19): 1793–803.
28. https://www.gilead.com/news-and-press/press-room/press-releases/2022/2/new-clinical-data-support-the-sustained-efficacy-of-longacting-lenacapavir-gileads-investigational-hiv1-capsid-inhibitor
29. https://www.ema.europa.eu/en/medicines/human/EPAR/sunlenca#

authorisation-details-section; https://www.thepharmaletter.com/article/uk-approval-for-gilead-hiv-drug-with-blockbuster-potential
30. https://www.fda.gov/news-events/press-announcements/fda-approves-new-hiv-drug-adults-limited-treatment-options
31. https://www.tht.org.uk/news/heterosexual-hiv-diagnoses-overtake-those-gay-men-first-time-decade
32. Vidal S.O., Beckerman E., Cihlar T. et al. 'Long-acting capsid inhibitor protects macaques from repeat SHIV challenges.' *Nature*. 2022 Jan; 601(7894): 612–6.
33. https://clinicaltrials.gov/study/NCT04994509?intr=Lenacapavir&rank=7; https://clinicaltrials.gov/study/NCT04925752?intr=Lenacapavir&rank=8

9: How to Innovate

1. Pao W., Nagel Y.A. 'Paradigms for the development of transformative medicines-lessons from the EGFR story.' *Annals of Oncology*. 2022 May; 33(5): 556–60.
2. 1,738 to be precise, according to the MHRA on 23 August 2023.

Glossary

(All websites accessed 1 October 2023 except where otherwise noted.)

1. Office of AIDS Research. Glossary of HIV/AIDS-related terms. US Department of Health and Human Services. 2021. https://clinicalinfo.hiv.gov/en/glossary/acquired-immunity.
2. National Cancer Institute. NCIthesaurus. 2023. US Department of Health and Human Services. https://ncithesaurus.nci.nih.gov/ncitbrowser/ConceptReport.jsp?dictionary=NCI_Thesaurus&ns=ncit&code=C154897
3. National Human Genome Research Institute. Talking Glossary of Genomic and Genetic Terms. US Department of Health and Human Services. 2023. https://www.genome.gov/genetics-glossary/Animal-Model.
4. NCIthesaurus. 2023. https://ncithesaurus.nci.nih.gov/ncitbrowser/ConceptReport.jsp?dictionary=NCI_Thesaurus&ns=ncit&code=C16504
5. NCIthesaurus. 2023. https://ncithesaurus.nci.nih.gov/ncitbrowser/ConceptReport.jsp?dictionary=NCI_Thesaurus&ns=ncit&code=C16295
6. NCIthesaurus. 2023. https://ncithesaurus.nci.nih.gov/ncitbrowser/ConceptReport.jsp?dictionary=NCI_Thesaurus&ns=ncit&code=C70913
7. NCIthesaurus. 2023. https://ncithesaurus.nci.nih.gov/ncitbrowser/ConceptReport.jsp?dictionary=NCI_Thesaurus&ns=ncit&code=C91459
8. US Food and Drug Administration. Breakthrough Therapy. US Department of Health and Human Services. 2018. https://www.fda.gov/patients/fast

-track-breakthrough-therapy-accelerated-approval-priority-review/breakthrough-therapy.
9. NCIthesaurus. 2023. https://ncithesaurus.nci.nih.gov/ncitbrowser/ConceptReport.jsp?dictionary=NCI_Thesaurus&ns=ncit&code=C21045
10. NCIthesaurus. 2023. https://ncithesaurus.nci.nih.gov/ncitbrowser/ConceptReport.jsp?dictionary=NCI_Thesaurus&ns=ncit&code=C200340
11. NCIthesaurus. 2023. https://ncithesaurus.nci.nih.gov/ncitbrowser/ConceptReport.jsp?dictionary=NCI_Thesaurus&ns=ncit&code=C16484
12. NCIthesaurus. 2023. https://ncithesaurus.nci.nih.gov/ncitbrowser/ConceptReport.jsp?dictionary=NCI_Thesaurus&ns=ncit&code=C17206
13. NCIthesaurus. 2023. https://ncithesaurus.nci.nih.gov/ncitbrowser/ConceptReport.jsp?dictionary=NCI_Thesaurus&ns=ncit&code=C15707
14. NCIthesaurus. 2023. https://ncithesaurus.nci.nih.gov/ncitbrowser/ConceptReport.jsp?dictionary=NCI_Thesaurus&ns=ncit&code=C13296
15. NCIthesaurus. 2023. https://ncithesaurus.nci.nih.gov/ncitbrowser/ConceptReport.jsp?dictionary=NCI_Thesaurus&ns=ncit&code=C206
16. NCIthesaurus. 2023. https://ncithesaurus.nci.nih.gov/ncitbrowser/ConceptReport.jsp?dictionary=NCI_Thesaurus&ns=ncit&code=C70676
17. NCIthesaurus. 2023. https://ncithesaurus.nci.nih.gov/ncitbrowser/ConceptReport.jsp?dictionary=NCI_Thesaurus&ns=ncit&code=C13231
18. NCIthesaurus. 2023. https://ncithesaurus.nci.nih.gov/ncitbrowser/ConceptReport.jsp?dictionary=NCI_Thesaurus&ns=ncit&code=C93020
19. NCIthesaurus. 2023. https://ncithesaurus.nci.nih.gov/ncitbrowser/ConceptReport.jsp?dictionary=NCI_Thesaurus&ns=ncit&code=C16700
20. NCIthesaurus. 2023. https://ncithesaurus.nci.nih.gov/ncitbrowser/ConceptReport.jsp?dictionary=NCI_Thesaurus&ns=ncit&code=C15263
21. NCIthesaurus. 2023. https://ncithesaurus.nci.nih.gov/ncitbrowser/ConceptReport.jsp?dictionary=NCI_Thesaurus&ns=ncit&code=C142585
22. NCIthesaurus. 2023. https://ncithesaurus.nci.nih.gov/ncitbrowser/ConceptReport.jsp?dictionary=NCI_Thesaurus&ns=ncit&code=C13249
23. NCIthesaurus. 2023. https://ncithesaurus.nci.nih.gov/ncitbrowser/ConceptReport.jsp?dictionary=NCI_Thesaurus&ns=ncit&code=C49135
24. Talking Glossary of Genomic and Genetic Terms. 2023. https://www.genome.gov/genetics-glossary/Non-Coding-DNA.
25. NCIthesaurus. 2023. https://ncithesaurus.nci.nih.gov/ncitbrowser/ConceptReport.jsp?dictionary=NCI_Thesaurus&ns=ncit&code=C22491
26. NCIthesaurus. 2023. https://ncithesaurus.nci.nih.gov/ncitbrowser/ConceptReport.jsp?dictionary=NCI_Thesaurus&ns=ncit&code=C16797
27. NCIthesaurus. 2023. https://ncithesaurus.nci.nih.gov/ncitbrowser/ConceptReport.jsp?dictionary=NCI_Thesaurus&ns=ncit&code=C135727
28. NCIthesaurus. 2023. https://ncithesaurus.nci.nih.gov/ncitbrowser/ConceptReport.jsp?dictionary=NCI_Thesaurus&ns=ncit&code=C86008
29. National Cancer Institute. Monoclonal antibodies. 2019. US Department of Health and Human Services. https://www.cancer.gov/about-cancer/treatment/types/immunotherapy/monoclonal-antibodies. Accessed 15 October 2023.

30. NCIthesaurus. 2023. https://ncithesaurus.nci.nih.gov/ncitbrowser/ConceptReport.jsp?dictionary=NCI_Thesaurus&ns=ncit&code=C16936
31. NCIthesaurus. 2023. https://ncithesaurus.nci.nih.gov/ncitbrowser/ConceptReport.jsp?dictionary=NCI_Thesaurus&ns=ncit&code=C18121
32. NCIthesaurus. 2023. https://ncithesaurus.nci.nih.gov/ncitbrowser/ConceptReport.jsp?dictionary=NCI_Thesaurus&ns=ncit&code=C70913
33. NCIthesaurus. 2023. https://ncithesaurus.nci.nih.gov/ncitbrowser/ConceptReport.jsp?dictionary=NCI_Thesaurus&ns=ncit&code=C187768
34. NCIthesaurus. 2023. https://ncithesaurus.nci.nih.gov/ncitbrowser/ConceptReport.jsp?dictionary=NCI_Thesaurus&ns=ncit&code=C15299
35. National Library of Medicine, National Center for Biotechnology Information. ClinicalTrials.gov Glossary Terms. 2023. https://clinicaltrials.gov/study-basics/glossary.
36. Ibid.
37. Ibid.
38. Ibid.
39. NCIthesaurus. 2023. https://ncithesaurus.nci.nih.gov/ncitbrowser/ConceptReport.jsp?dictionary=NCI_Thesaurus&ns=ncit&code=C16983
40. NCIthesaurus. 2023. https://ncithesaurus.nci.nih.gov/ncitbrowser/ConceptReport.jsp?dictionary=NCI_Thesaurus&ns=ncit&code=C48674
41. NCIthesaurus. 2023. https://ncithesaurus.nci.nih.gov/ncitbrowser/ConceptReport.jsp?dictionary=NCI_Thesaurus&ns=ncit&code=C783
42. NCIthesaurus. 2023. https://ncithesaurus.nci.nih.gov/ncitbrowser/ConceptReport.jsp?dictionary=NCI_Thesaurus&ns=ncit&code=C16700
43. NCIthesaurus. 2023. https://ncithesaurus.nci.nih.gov/ncitbrowser/ConceptReport.jsp?dictionary=NCI_Thesaurus&ns=ncit&code=C14353
44. NCIthesaurus. 2023. https://ncithesaurus.nci.nih.gov/ncitbrowser/ConceptReport.jsp?dictionary=NCI_Thesaurus&ns=ncit&code=C17106
45. NCIthesaurus. 2023. https://ncithesaurus.nci.nih.gov/ncitbrowser/ConceptReport.jsp?dictionary=NCI_Thesaurus&ns=ncit&code=C18279
46. NCIthesaurus. 2023. https://ncithesaurus.nci.nih.gov/ncitbrowser/ConceptReport.jsp?dictionary=NCI_Thesaurus&ns=ncit&code=C17206
47. NCIthesaurus. 2023. https://ncithesaurus.nci.nih.gov/ncitbrowser/ConceptReport.jsp?dictionary=NCI_Thesaurus&ns=ncit&code=C17208
48. NCIthesaurus. 2023. https://ncithesaurus.nci.nih.gov/ncitbrowser/ConceptReport.jsp?dictionary=NCI_Thesaurus&ns=ncit&code=C17207
49. NCIthesaurus. 2023. https://ncithesaurus.nci.nih.gov/ncitbrowser/ConceptReport.jsp?dictionary=NCI_Thesaurus&ns=ncit&code=C14184
50. NCIthesaurus. 2023. https://ncithesaurus.nci.nih.gov/ncitbrowser/ConceptReport.jsp?dictionary=NCI_Thesaurus&ns=ncit&code=C19108
51. NCIthesaurus. 2023. https://ncithesaurus.nci.nih.gov/ncitbrowser/ConceptReport.jsp?dictionary=NCI_Thesaurus&ns=ncit&code=C62195
52. NCIthesaurus. 2023. https://ncithesaurus.nci.nih.gov/ncitbrowser/ConceptReport.jsp?dictionary=NCI_Thesaurus&ns=ncit&code=C17672.

INDEX